María Eugenia Culla

Digital resources in mathematics teaching

María Eugenia Culla

Digital resources in mathematics teaching

Implementation of a virtual teaching-learning space for mathematics in the fourth year of high school

ScienciaScripts

Imprint

Any brand names and product names mentioned in this book are subject to trademark, brand or patent protection and are trademarks or registered trademarks of their respective holders. The use of brand names, product names, common names, trade names, product descriptions etc. even without a particular marking in this work is in no way to be construed to mean that such names may be regarded as unrestricted in respect of trademark and brand protection legislation and could thus be used by anyone.

Cover image: www.ingimage.com

This book is a translation from the original published under ISBN 978-613-9-40772-9.

Publisher:
Sciencia Scripts
is a trademark of
Dodo Books Indian Ocean Ltd. and OmniScriptum S.R.L publishing group

120 High Road, East Finchley, London, N2 9ED, United Kingdom
Str. Armeneasca 28/1, office 1, Chisinau MD-2012, Republic of Moldova, Europe
Printed at: see last page
ISBN: 978-620-8-05346-8

ACKNOWLEDGMENTS

Thanks to the tutors of each course for their patience, dedication and motivation, which made the study of these years much easier with their guidance and help.

Special thanks to Mg. Rubén Pizarro for his support in this last stage by accepting to be my director in this work.

Special thanks to the principal of "Edgar O. J. Morisoli" High School, Professor María de los Ángeles Pereyra, for allowing me to do the necessary internships at the institution, as well as to the students and teachers who collaborated whenever I needed, sincerely thank you very much.

And above all, thanks to my family for their support and understanding, especially to my daughters and my husband for their unconditional support during these years of study.

SUMMARY

The objective of this work was to design and then implement a Virtual Teaching and Learning Space (EVEA) that allows students to learn mathematical knowledge through the appropriate use of different technological resources. This proposal is aimed at students in the 4th year of the morning shift, who, in a survey conducted on the use of ICT in the classroom, the teachers of this course stated that we carry out classroom proposals with technological resources. The students, on the other hand, considered that such proposals were not attractive, assuming that they have the use of devices in the classroom, but not for educational purposes. For this reason, an EVEA is developed specifically in the area of mathematics where the aim is to teach with technology, by means of an easily accessible platform such as Classroom. Knowledge is organized by topics, interactive resources are provided where students, through ICT and virtual classes, are expected to acquire autonomy and commitment in their work, giving an account of the concepts learned through the different activities proposed. In this way, the students are expected to play an active role, the teacher being a mediator of knowledge and a moderator in the proposed online classes. Facilitating the ICT resources to be used according to the Priority Learning Nuclei proposed by the Ministry of Education in conjunction with the interests of students, designing activities that allow the construction of meaningful knowledge.

KEYWORDS:
Mathematics- ICT-EVEA- Resources

Table of Contents

JUSTIFICATION/ DIAGNOSIS

It is important that as teachers we assume that new technologies influence the actors in the educational process and have effects on individuals both in the social and work environment. Thus, we must prepare our students for the use of new resources and for this to happen we have the responsibility to incorporate Information and Communication Technologies (ICT) in the teaching-learning process as a didactic and pedagogical tool.

It is observed that there is currently a very frequent use of cell phones by students in all areas of their daily lives, even in the classroom, but it is not used as a resource that serves for the organization and learning in the different subjects offered by the ministerial curriculum. There are very few students who make proper use of new technologies and it is necessary to work on the use or abuse of them as they can cause problems in the short or long term in terms not only of school performance, but also disciplinary problems, alterations in neurological development, etc.

Considering that students master different social networks and create content using different resources and information, it is necessary that ICT be incorporated through these devices not only as technological instruments, but also as a didactic tool through virtual teaching-learning environments (EVEA). Promoting the use of software specific to the subject and different web resources that can be proposed by the teacher, and also by the students according to their interests and interpretations.

The use of these environments not only allows the management and storage of information, but also has its own tools that enable a personal and collaborative learning process, providing spaces to authorize, communicate and discuss, giving independence among the actors of the educational community. They also favor the motivation and interest of the youngest thanks to animation, videos and exercises that are eye-catching and fun, thus promoting a teaching model based on interaction and the development of creativity.

This work was planned for students in the 4th year of the orientation cycle, who attend the Secondary School "Edgar O. J. Morisoli" in the city of Santa Rosa, Province of La Pampa. The institution has internet access with good performance, was included in the Plan Conectar Igualdad with its corresponding designated Referent and in 2017 was incorporated into the National Integral Plan for Digital Education (PLANIED)[1] .

These programs proposed by educational policies at the national level allow students to have better access to digital technologies during school hours, but as Burin et al. (2016) state, having access to digital does not mean that they have better learning, nor that technological tools alone innovate in the classroom, for this to happen, changes in the mentality and sociocultural practices of the main actors (teachers/students) are needed/required. A cognitive domain is needed, such as search and navigation skills, being necessary to organize sites and pages so that students can understand the proposed contents, stipulating times and activities, as well as the search for interactive resources that allow to account for the knowledge acquired by the students.

Due to the importance of pedagogically planning the Mathematics curricular space, it is appropriate to use Classroom as a virtual platform, which besides being a free service, is an educational social network that allows a dynamic approach, ease of navigation, agility in the qualification of participating students in a safe and controlled environment. It provides communication and organization of the contents proposed in each class. This platform provides an easy configuration, saves time and paper, in 2020 (due to the demand due to the Pandemic situation) it improved its communication and comments system, has no ads and works very well with Google Forms, Calendar, Gmail and Drive.

As mentioned above, considering that these devices can be used as didactic tools within the educational environment, and through proposed actions, students should have access to information individually or collectively, inside or outside the school environment, knowing how to identify the veracity of different sources, achieving an autonomous and responsible role. In this way, by making good use of computer resources, they can acquire new knowledge and go deeper into what interests them most, being able to

---

[1] https://siteal.iiep.unesco.org/bdnp/43/resolucion-1536-e2017-plan-nacional-educacion-digitalplanied

select and value the information they consider relevant.

It is important the fact that the use of ICT will allow a more dynamic process, promoting flexibility in time and space, as well as contributing to collaborative and constructivist learning, the search and publication of different resources that allow to account for the knowledge acquired from the curricular area. It also facilitates not only face-to-face learning but also blended and on-line learning.

Although the last two mentioned modalities are usually found in higher education, tertiary education or job training, in some opportunities, the Ministry of Education of La Pampa proposes flexible trajectories for students with certain characteristics; which promotes that students should perform in different virtual communities through the different stages and situations of their lives, being able to access information, knowledge, access and job training at any time and place. This means that a wide range of different models of educational practices should be considered for the different needs and requirements of our young people, whose lives go through processes of change in various ways (physical, cognitive, emotional, etc.).

It is essential to teach with technology, implementing different software of the curricular space, as well as multimedia resources that allow the development of creative didactic proposals where the student is active, critical and responsible in the teaching-learning process. In this way, students are expected to have a constant communication with their peers and with the teacher, being able to promote debates and defend their positions in the classroom about the knowledge proposed by the subject in each class.

There are teachers with fear, resistance and frustration to the use of Information and Communication Technologies or simply in some cases they do not want to leave their comfort zone, being traditional when teaching. This means that we must promote change in the future practices of teachers, who should no longer have the role of transmitting knowledge, but should be the guide and mediator of information, with the ability to adapt to cultural changes related to ICT (Area and Adell, 2009).

New technologies go beyond knowing how to use them, we must really know about their applicability, because from there we can be more creative and innovative, putting into practice new educational models, considering the needs and concerns of a student and institutional population, allowing other forms of participation. For the above mentioned, the different interests of the students should be contemplated and respected, stimulating them in the construction of knowledge and critical thinking, for this reason it is promoted to review and reflect on the current teaching practice so that educational activities with ICT should be generated to be implemented in the classroom.

At the intermediate level, the prepared teacher must provide tools that allow the student to analyze, compare resources, know how to communicate, obtain information, select it and organize it, promoting the interest of the students to improve the quality of learning.

For this transformation to occur, knowledge must be planned and created based on the use of ICT, stimulating students' cognitive knowledge by using problem situations and why not starting from erroneous situations (remember that mistakes are usually the fundamental basis for learning and understanding). That is to say, creating resources and strategies knowing how they learn and what they understand in order to plan learning styles and needs, knowing the technological resources to give them a good didactic use, promoting a meaningful experiential learning. Evaluating in a permanent way, being able to detect tensions from the beginning, allowing constant feedback and informed decision making by teachers and students, and in this way giving rise to an evaluation in process for learning.

When implementing ICT, the teacher should not only evaluate the process, but also the constant expressions of the students, the capacities to expand, to integrate concepts and to incorporate different sources of information to those already provided in the different situations proposed for learning (Cobo, 2016).

The use of new technologies also brings about an important change at the institutional level by modifying the methodological order and the teaching of subjects, proposing didactic sequences integrated by new technologies that are transversal to all curricular areas. Thus, different teaching modalities emerge, which we must constantly adapt according to the social, cultural and working world needs for which the new generations of students must be trained.

In order for ICT to be incorporated in the educational environment, certain limitations must be overcome, one of the most important of which is expressed by the following authors:

> Within the limitations identified, special interest has been given to the development of training and capacity building actions in educational institutions and organizations. Among the responses to these needs, the incorporation of the e-learning and b-learning modality to the training processes in organizations that are mainly based on open and distance education stands out (Montoya ALA, Parra CMR, Lescay AM, 2019, pp 246.).

Finally let's keep in mind that all students in one way or another have acquired "digital skills" but from the educational field we must teach them cognitive skills such as: knowing how to search for information, knowing different platforms, performing dynamic navigations, producing knowledge and allowing not to have a single view of things (Adell, 2018).

2. PROBLEM STATEMENT

In the subject "Practice I" of the Master's Degree in Teaching in Digital Scenarios, a field study was carried out in order to find out how teachers of certain curricular areas implement and teach the use of ICT in their practices. We surveyed 3rd and 4th year students and teachers of Mathematics, Information and Communication Technology and Technological Processes. The middle school shift in which the survey was conducted has 273 students and it was about the use of technological devices and resources. The data collected show that students use ICT in the classroom (either with their own devices or those provided by the institution) in most cases to use social networks such as: Facebook, Instagram, Twitter, Snapchat, etc. Many of the students play online, but not on their cell phones because they claim that storage is sometimes not enough.

In these situations, distractions in class work and low academic performance of students are observed. The challenge is to work so that as teachers we can elaborate interesting didactic proposals using ICT and so that students learn to use these resources to acquire knowledge. One of the results of the data collected is that a large percentage of students use cell phones more than computers, that all students have mobile devices, and that the proposals made by teachers to use ICT are indifferent to them.

We consider that one of the problems is that ICTs are not implemented in such a way as to take advantage of their full potential in the teaching-learning process and are only used as a bibliographic repository in some cases. For this reason, they are sometimes seen as a distraction factor since they are not considered by students as a tool for their continuing education. Therefore, we must rethink as teachers the didactic proposals offered to incorporate technology in the classroom and make good use of technological resources for pedagogical purposes.

3. OBJECTIVES

3.1 General Objective

Design and implement a virtual teaching and learning environment for mathematics at the intermediate level.

3.1.2 Specific Objectives.

- Select audiovisual, interactive and simulator materials.

- Design work activities that contemplate the knowledge proposed from the NAP (Priority Learning Nuclei), adapted to the virtual teaching and learning environment.

- Elaborate follow-up proposals for the virtual classroom to ensure that students understand the proposed knowledge.

- To promote conceptual schemes that allow the understanding of knowledge for those who enter the virtual classroom.

4. THEORETICAL FRAMEWORK

4.1 Connectivism and Collaborative Learning

Nowadays, where globalization is present, different learning theories are being studied based on some pre-existing ones, thus giving rise to a new theory of the digital era called Connectivism. It is of utmost importance since according to Siemens (quoted by Gutiérrez, L, 2012) educational institutions become part of the economic market providing knowledge and information services as a competitive product.

It can be said that learning must be connected and that the connection is much richer if the work is shared with others and facilitated with technologies. While some authors doubt that Connectivism is a learning theory, it can be incorporated into existing ones.

The connectivist proposal has some issues to be taken into account. Connections are essential in the learning process since it assumes that inquiring and researching in networks provides knowledge, but does not guarantee learning. Learning is considered as an immediate experience where the advantage is a collaborative management in editing, organizing and retrieving information.

Another issue is the deinstitutionalization of education and forgetting the instructional design, the idea is to sustain that one can learn at any time and place but always mediated, where the teacher will rethink the knowledge of the curricular programs, allowing the self-correction of the students' learning, intervening and mediating critical learning in the networks and providing them with personalized environments that favor meaningful learning in the students. In this sense, the main protagonist is collaborative learning where interaction and interactivity contribute to active learning, providing significant support among peers and an important role of the teacher as mediator.

Siemmens is a great proponent of Connectivism and argues that ICTs alter our way of living, since interactive tools allow us to manage information, generating an active, critical and fast thinking. Connectivism responds to the demand of the XXI century education that is invaded by new technologies, the digital era has great advantages such as the incursion into different areas of knowledge, the possibility of acquiring knowledge in different ways and continuously.

With respect to learning, it has the advantage of being collaborative (as mentioned above) because ICTs provide tools that facilitate group work, making it possible to address different areas in a cross-cutting manner, since the same topic can be dealt with a wide range of information, promoting digital literacy. The disadvantage is that, if society is not used to working collaboratively, we cannot affirm that learning is effective, as well as the cost of technology and the constant advance of the same with respect to software and hardware. As far as students are concerned, it allows them to access information at any time and place, work in teams promoting interest and developing skills in the search for information. A setback is the amount of information material they obtain, which often leads them to end up in a "cut and paste"; that the work team is not conducive to collaborative work and non-academic distraction on the web.

The teacher who acquires a wide knowledge in technologies must promote collaborative learning and know how to take advantage of resources such as videos, simulators and different internet pages. Continuous investment and training are always needed.

As previously mentioned, due to globalization and needs, education is becoming part of the market and according to Morrian (cited by Gutierrez, 2012) students are going from being learners to being considered consumers. It is a fact that all educational programs incorporate ICT as a fundamental tool for learning, since technology is part of today's economy. Learning in networking is the great difference between Connectivism and the other learning theories (behaviorism, constructivism and cognitivism), it also allows self-organization, so learning is self-organized, promotes learning open to information and capable of classifying its own interaction with the environment.

In Connectivism the network is learning, in the abundance of knowledge that characterizes today's society, allowing different points of view and opinions that lead to experiences that allow better decisions. It is presented as a pedagogical proposal that provides the ability to connect to each other through social networks and/or collaborative tools, extending practices beyond the classroom by providing real life experiences, so it is essential to take into account the needs of those who learn, the student explores the objectives and these should be defined by themselves as well as the resources to use and thus motivate them to learn. The essential tools are synchronous and asynchronous, and belong to the web 2.0 where users are more active when receiving information and in the collaborative creation of contents.

4.2 Open Learning and PLE

Another theory based on the use of technologies is the theory of Open Learning. Everything mentioned so far leads us to a new paradigm: Techno-economic, since world economies are constituted in research, patenting, agreements with companies and universities, always depending on the context in which the actors are according to conditions: social, political and economic.

According to Salinas, J. (2013), open learning or also called open education, is an education for all where access to: open education programs that grant national degrees, open access courses from formal and non-formal education, open educational resources, online books as well as research papers and open data.

Open education is an objective of educational policies that is characterized by being available to all, overcoming barriers to the diversity of students that may exist, this proposes an open and flexible learning. The open condition is related to the use of technology, since it does not deny access to anyone, thus using the available ICT.

Nowadays society manages social networks and is immersed in connectivism, where learning does not only take place in institutions, therefore, we speak of a flexible and open education in which each subject develops his PLE (Personal Learning Environment), which is understood as an open learning where the user accesses and controls the way he learns, in this way the didactic process is centered on the learner. The PLE are the individual educational platforms of each student in order to direct their own learning and achieve educational objectives, where the system adapts to the student integrating non-formal and formal learning. It can be presented in two ways: technological and pedagogical, it is based on the use of ICT in the learning process. As we mentioned before, the PLE is part of open learning, and the latter has two dimensions; one is related to the administrative, i.e. the organizational aspect regarding how, where and with what technological resources, while the other dimension is related to the didactic aspect, i.e. goals, sequences, teaching strategies, etc.

Open learning is centered on the learner who makes decisions about whether or not to carry out the proposed activities, selects content, and also how, where and when to learn, who to use, etc. The teaching-learning methods use appropriate ICTs in a networked environment, allowing the learner to create and at the same time consume information and knowledge, accessing a new way of learning. The EVEAs must generate tempting proposals that allow expanding and motivating the need to generate new knowledge, which implies a new look at the existing pedagogical models.

Today we can speak of interactive communities between students and teachers, and this is due to the fact that sometimes communication is mediated by a computer, but the potential for achieving self-directed learning lies in the didactic design with which we are going to work on the subject in question and not in the ICT.

ICT open different sources of change when considering educational technologies such as changes in the conception of how the classroom works, didactic processes, basic resources, materials, access to networks, manipulation of information and see the cost benefit with respect to didactics. Some of the objectives obtained by incorporating technologies in education is that they allow building solutions to individual / social needs of the learner and improve the quality and effectiveness of interaction.

4.3 Information Society and Knowledge Society

The presence of technology, both in the economy and in the productive apparatus, has an impact on social relations and makes all members of society obtain and share information at any time and from any place, giving rise to the information society. It is based on new educational models that are based on the technological infrastructure to process and transmit information, facilitating the activities of millions of people around the world with the creation and interaction of electronic content, promoting lifelong learning processes that allow modifying different work habits and thus promoting different ways to successfully face the challenges of the present and the future. The Information Society is the sustenance and support of the Knowledge Society, that is to say, it is a previous stage where knowledge is the engine to drive innovation in the economic system.

On the other hand, the concept of knowledge society arises because knowledge is fundamental as a source of wealth production, since it is based on the production of services, an essential link in the social processes of the different areas, this being an economic resource. With the advent of ICTs, the new activities of the global market consisted of generating, storing, distributing and processing information.

Knowledge is necessary for the productive world and the global economy, because the new productive strategies are based on the use of information for business purposes, so it seeks to study educational models that train qualified people for companies; understanding the knowledge society as knowledge economy.

In the advance of globalization in which we are immersed and in which we participate, ICTs allow us to combine all types of mass communication, and because students are not exempt from what happens in society, we must think of a new student profile that cannot be taught in the traditional way, considering them as mere receivers. One of the pillars of the Knowledge Society is the access to information and freedom of expression where knowledge is communicated and shared, giving great importance to education and access to technology in order to form competent citizens in the globalized world, with an intellectual preparation to perform effectively in a digital society, where knowledge is transformed into a main tool for their own benefit.

Both science and technology contribute to the Knowledge Society since their main objective is to build knowledge through scientific-technological progress in educational institutions. In this way we observe that some factors are innovation and technology (which is developing faster and faster). Thus, innovation, technology and creativity are inseparable within the Information Society and the Knowledge Society. Therefore, the new demands in education so that the changes are truly innovative is to contemplate an interdisciplinary work that integrates not only technological knowledge, but also pedagogical, promoting the training processes not only obtaining results from the educational performance, but in the construction of knowledge of the different areas of knowledge through technological incorporation.

With respect to the knowledge and information societies, both exist thanks to ICTs, since they facilitate communication and the massive exchange of information, thus giving rise to the generation and transmission of knowledge. Within the new technologies are the virtual platforms, which contribute to innovation in education leading to the implementation of digital libraries, and educational devices that are available to everyone for free, allowing the knowledge society to prepare for digital literacy, accessing the new educational models to adapt and respond to new technologies that appear.

In this way we ensure that the student must acquire digital skills, that is to say that we must make them digitally literate so that they feel included in today's world, thus reducing the digital divide, and for this reason we must evolve both in education and training, as well as in teaching practices. That is to say, practices must be adapted according to the advances, to the inclusion of technologies in education and to the interest of the students, so that knowledge is not only built individually, but also promotes its collective and cooperative construction, distributing it in real time (Castell, 2009), otherwise it causes reluctance and frustration in the classroom.

In order to reduce the aforementioned gap, socioeconomic factors must be taken into account, and it is not enough to have one computer per student-teacher, but also to have the necessary infrastructure and information technology to carry out the programs proposed by the institutions and homes.

The importance acquired by online education in 2020, when the Pandemic was declared, will mark a before and after in pedagogical practices, highlighting the social, cultural and economic gaps mentioned above. The challenge is undoubtedly to maintain education and promote meaningful learning with the challenge we have as teachers to generate our own strategies and learning to work in virtual environments teaching our students to manage them, generating a communication scenario with technologies that were previously only used as support and become the main tool providing a new communication scenario as far as education is concerned.

We know that in education there is a constant activity of adaptations to new social, cultural and technological paradigms. The pandemic situation presents a new reason to modernize the processes used in the educational system, where students and teachers had a great challenge to move from a face-to-face context to a virtual one, when physically closing the educational institutions exposes the need to innovate in the teaching-learning processes.

4.4 ICT, Education and Digital Divide

At the secondary level, there was no history of virtual courses prior to the 2020 academic year, as was the case at the university level and in distance education, which was only implemented in the EPJA modality[2] (Continuing Education for Youth and Adults) and consisted of presenting work in person with some classes scheduled for consultations.

At first, ICTs are implemented at the secondary level through the National Program "Conectar Igualdad" in face-to-face classes, providing flexibility and trust between students and tutors, where it is important to analyze the triad: teacher, students and contents to be addressed, since the teacher is often overwhelmed in the virtual world with the number of students in his charge and the time involved in correcting and following up the teaching process. With respect to the activities that are rich in discussion through forums, wikis, shared documents, we can maintain that the teacher, being the mediator, has the same functions that he/she has in the physical classroom. Online classes through Zoom or Meet, allow an important parallelism with

---

[2]

http://digesto.tcuentaslp.gob.ar/digesto%20tribunal/Otros%20Organismos/Mrio.
%20de%20Educacion/Resolucion%20112-2018%20-%20Anexo.pdf

face-to-face classes where the teacher must manage the meeting and provide feedback to students through questions.

If instead of proposing synchronous classes, teachers give their classes through video tutorials and explanatory videos on a given topic, these, being asynchronous, are more attractive than reading texts. Of course, the key points of virtuality are the teachers, the students and the proposed contents, it is important to know what expectations, preparation and participation the students have, while teachers must understand and understand the new role, the management and the way of teaching that we have in the digital era by proposing multimedia contents.

Of course, educational technology is based on other theories that contribute to its development, some of them are: the pedagogical theory that contributes from didactics, institutional organization and curriculum development; communication theory since education is considered as a communication process contributing concepts and instruments to educational technology; the general theory of systems and cybernetics where a process of instructions is designed that must contemplate objectives, methods, resources where all members of the educational community participate. But how are learning theories and computer theories related, there are educational software and the possible implementation of these in the classroom leads to a restructuring of teaching strategies and their objectives.

There are applications that, due to their configuration, make the student play a passive role. In order for this not to happen, educational programs must provide activities and games that allow the understanding and construction of knowledge.

The theory of situated knowledge holds that the Internet is a medium for learning, where the learner obtains knowledge in a situated way, learning through perception. In this context, the Internet responds to the premises of the aforementioned theory with realism and complexity, allowing real exchanges between users from different cultural contexts but who have similar interests.

In order to generate a quality virtual education, adequate technological resources are needed, access to different educational programs that allow to generate effective learning by creating satisfactory environments for teachers and students. In principle, the main use was given to social networks making it a highly valued resource, with them virtual environments are generated that allow in a simple way to maintain communication with students, such as WhatsApp and Facebook.

Let us remember that there is a virtual or digital divide that affects the online modality since not everyone has access to information and communication technologies, thus deepening the socio-educational inequalities that were exposed in the 2020-2021 school year due to the situation that was experienced worldwide with respect to the pandemic.

A consequence of narrowing the digital divide is that we can also narrow the cognitive divide, since by allowing everyone to have access to computers and use of the Internet, we can access knowledge and knowledge through information by making use of technology. For this to happen, schools are the passport for our students to learn how to use computers and information, playing an active and participatory role in the construction of knowledge, and for teachers to improve their pedagogical and didactic practices.

We assume that including ICT in education guarantees education for all, because through the network information is acquired by everyone regardless of where they are, but in reality not everyone has access to the Internet, so there is discrimination against those who cannot have access to these new tools because of their economic resources or the place where they live. In this way, a Digital Divide is created, and it is defined as the inequality of possibilities that exist to access information, knowledge and education through new technologies. This technological marginalization is transformed into social and personal marginalization, therefore, this is how the Social Divide manifests itself.

Nowadays, Internet connection is almost total worldwide, where access to ICT in each country is linked to the economic conditions of the country, and thus, as progress is made in technical and economic levels, social aspects often lag behind. Let us bear in mind that ICTs in a global economy are a strategic and competitive element, so it may happen that what is free access today may have a cost tomorrow, implying that it will probably not be available to everyone. Although at some point everyone will have access to new technologies, in the meantime the differences will be accentuated and then it will be more difficult to come closer. On the other hand, there are reports by the International Telecommunication Union that state that ICTs can: reduce poverty (as it allows access to commercial information and the global market to developing countries and companies), improve health (facilitates the monitoring and exchange of information on patients and diseases, as well as the advancement of different global research on diseases), promote sustainable developments (technological resources and software that allow optimizing progress to promote environmental care) and enrich education and inclusion (as ICT improve teacher training by allowing interaction between colleagues, accessing a wide variety of material and resources, etc.).)

According to Cabero Almenara (2014) when talking about the Digital Divide there are two lines, the soft one where the problem to be solved is only infrastructure, technology, telecommunications and information technology and the hard one which is the most complex one where the Divide is a consequence of social and economic inequality of capitalist society. At the time of proposing solutions, one seems that just by accessing the Internet all the problems related to learning are solved, but from the other perspective, as the digital divide is a consequence of inequality, either we solve it or, no matter how much access to networks and platforms is fixed, only an exclusive part of society has access to it.

Of course, the digital divide is not only between countries, but also exists within them. It is not enough for everyone to have a computer, but they must be digitally literate, this means that we must have skills for work, community and social life where we must know how to handle information and have the ability to assess the relevance and veracity of the same on the Internet.

Literacy must create competent people in three aspects: knowing how to handle ICTs instrumentally, having positive and realistic attitudes in their use, and knowing how to evaluate their messages and needs in their use. It should also provide information necessary to recognize values and interpret different systems of social, communicational and cultural organization.

The students of the future must obtain competencies such as being able to adapt to rapidly changing environments, work collaboratively, have creativity to solve problems, learn knowledge, assimilate new ideas, take initiative and be independent, identify problems and develop practical and alternative solutions, gather and organize facts, make systematic comparisons, etc. All of this leads to new competencies to know how to interact with information, to intellectually handle different systems and to know how to work with different technologies.

Cabero Almenara (2014) argues that currently much importance is given to the inclusion of ICT in the teaching-learning process, so much so that there are concepts that are changing, for example, the computer classroom became the computer in the classroom and being on the network became part of the network. On the other hand, something to keep in mind is that having equal access to knowledge is not the same as being equal to knowledge, this means that when surfing the web from one side to the other we must perform significant cognitive actions in order to find the information required according to the proposed or posed situation.

An important capacity that exists in the digital divide is the language gap, since most of the new resources or web sites on the network are in English, the generation gap is also important and has a great impact on education, since students have a remarkable command of ICT communication in cybersociety and teachers feel insecure in the face of the technological framework. This is why, as mentioned in this paper, teacher training programs should be changed to include ICT in the curriculum.

4.4 Inclusion of ICTs in the Teaching-Learning Process and in the Curriculum

It is necessary to think about new Virtual Teaching and Learning Environments (VLEE), to investigate the different online resources available and to plan pedagogically in virtual classrooms. With regard to this new way of implementing teaching practices, we must take into account four dimensions: Informative (the materials that are presented to the students so that they can work autonomously), Praxical (activities and actions that the teacher plans for the student to carry out in the virtual classroom), Communicative (the resources for interaction and communication between teachers and students) and Tutorial and educational (the teacher follows, accompanies, motivates, etc. In this way we will be promoting an e-learning modality where information can be accessed individually or collectively from different sources and devices, giving the student an autonomous and responsible role, with flexibility in time and space, promoting collaborative learning and being an exhibitor of contents.

E-learning can be defined as an electronic lesson that allows the acquisition of knowledge through the Web. In the e-learning training scenario there are different scenarios: face-to-face (traditional education), blended learning (face-to-face combined with teacher tutorials), distance learning (when the didactic material is sent electronically to the students), blended learning (student-teacher/student online, including face-to-face activities) and elearning (online student with tutorial organization).

Fernández Tilve et al. (2013) argue that an important challenge of e-learning is to create content both in terms of material and layout, also taking into account some dimensions such as: technical, aesthetic, didactic, etc. In this sense, it is a task for the e-learner in which the accumulation of information is transformed into knowledge.

To speak of e-learning then is to speak of a knowledge-based economy, based on the information and knowledge society for all, where it is not only important what ICT resources the actors of the educational system use, but also what competencies they acquire in the virtual classroom that is proposed to them, which should become an important and powerful didactic tool that allows diversifying activities and learning experiences where ICT are efficiently and effectively integrated into the curriculum.

The model mentioned in the previous paragraphs allows organizing classes for students in different geographical locations, allows students to acquire the material at any time and it is interesting to propose that the material be interactive and autonomous, offering "face-to-face" activities, thus employing different approaches.

When preparing an e-learning course, it is important to be clear about the objective, where it is also important to have information about the students and thus be able to propose contents that are of interest to them, the result will surely be an effective course. Obviously, in addition to the students' interest, for a course to be effective it is necessary to take into account the resources and the technological limitations of the students. With respect to the pedagogical methods that can be implemented in e-learning, there are: interactive lessons, simulations, online discussions, evaluation proposals and surveys.

The inclusion of ICT promotes for example the possibility that two or more people at the same time can perform the same practical work with technology as a mediator of the process, allowing students to produce collaborative learning, and also opens the doors for change in all areas of people, but they do not have a direct effect on learning, to have an impact it is necessary to propose planned activities, they should be used as a support resource, acquire and develop skills to handle ICT and not plan parallel to the teaching process.

Both the knowledge society and new technologies affect all levels of the educational system, schools should bring students closer to the current culture, where educational activities should be aimed at psychomotor, cognitive, emotional and social development. In this sense, ICT are the means of exposure and instrument to process information through different software, a variety of didactic tools for learning as well as for self-evaluation; they are also open sources of information through all websites and platforms; they are a channel of face-to-face communication through blackboards as well as a channel of virtual communication through applications for forums, videoconferences, etc.

In this sense, it would be convenient for institutions to have a training plan to integrate new technologies in education, where the purpose of the plan would be: to give everyone access to information and communication technologies, to promote the use of the Internet and motivation, to strengthen the development of networks and cooperation among the different actors, to create an open, innovative and attractive learning environment.

When ICTs are included in a school, three scenarios can be presented: the technocratic scenario when technology is included to improve the information process and then as a source of information and provider of didactic material; the reformist scenario that consists not only in including ICTs as a source of information, but also in implementing new teaching-learning methods; and the third and last scenario is the holistic one where a profound restructuring of all the elements is carried out (Alcántara Trapero, 2009).

As everything that is implemented, ICTs have advantages and with respect to these, from learning, new technologies promote motivation, interaction and with them the development of initiatives, learning through error as a source of learning and greater communication between teachers and students. They encourage collaborative learning and the possibility of working transversally, promote digital literacy to develop the ability to search and select, as well as to visualize simulators.

Regarding the disadvantages, in learning it is necessary to take care of the distraction and dispersion of students when surfing websites that are not conducive to study, often causes loss of time searching for a lot of information, which must be selected according to the veracity of the same and that are not decontextualized with respect to the proposed learning situation. If students do not make good use of ICT, they may suffer addiction, isolation, visual fatigue and bad posture when they spend a long time in front of a device. The disadvantage for teachers is stress, since it consumes too much time in the preparation and search for appropriate didactic resources.

Regarding the role of the teacher, he/she will no longer be the one who knows everything and transmits knowledge, but a mediator who must motivate, reinforce and guide the practices of the study habit by strategically planning virtual classrooms to capture the attention of our students and promote meaningful learning based on the construction of knowledge. Teachers should use their ICT skills to include them in the development of their practices, look for resources and materials that allow them to work formally and informally, face-to-face or virtually. Design technological environments and materials based on the students' interests.

For this to happen, the teacher must analyze the variety of media that can be adjusted and enhance to the maximum the discipline in which he/she works;

for example, in the text "Teaching and innovations in the classroom for the new century" (1997) Cecilia Cerrotta states "[...] To have access to a computer (at least) and to the diversity of existing products that can be enhanced in favor of the disciplines taught in middle school" (p. 11). (p. 11). It is necessary to invite to think and reflect on proposals and activities to work with students about technology and its impact on the social, cultural and political life of society.

On the other hand, it is important to include proposals mediated by digital technologies in curriculum design and development. The curriculum, when considered as an educational project, is a political-educational proposal that must be developed and interpreted according to the context. It must respect the different cultures, customs and address the realities to which they aspire and have, i.e. its purpose is to improve the quality of life of students.

All schools have an IEP[3] (Institutional Educational Project) which is developed in the following areas
Institutional meetings with all teachers in each area where knowledge and teaching and evaluation methodologies are discussed. It is here where we must begin to include ICTs so that they can be used for curricular purposes, supporting the disciplines and stimulating learning.

There are six models to include ICT in the curriculum: Nested, which is when the work of diverse skills is stimulated; Woven, in this model a theme is proposed and woven with other contents, therefore, ICT serve as support to examine information; Threaded, as the word indicates, it is about threading social skills of thinking, intelligence and study by means of diverse disciplines through the search for overlapping ideas and concepts using ICT; Immersed when one can filter contents with the use of new technologies on a particular discipline; and Networked when the student filters his learning and generates internal connections that lead him to interact with external

---

[3] The PEI is the general statement that concretizes the mission and links it to the institutional development plan, which gives meaning to short, medium and long term planning.

networks using ICT. To be able to include technologies, one must first overcome fear and discover the potential they have, then one must know them and use them in various activities and finally include them in the curriculum for an educational purpose, the incorporation of course must be gradual, starting with guided and programmed activities to get to know the technological resources up to the use to create constructivist learning environments (Orjuela Forero, 2010).

The first step in integrating ICTs into the curriculum is to diagnose what students and teachers know about them and what their needs are, and then to train all the actors in the teaching-learning process based on the diagnosis, plan how ICTs will be integrated, specifying methodological and pedagogical strategies, resources and tools, and finally develop the planned strategies, thus beginning to recognize the potential of ICTs in the pedagogical process and freeing the fear of implementing them. The use of ICT is hypermedial, so learning by different means arises to respond to the diversity of multimedia learning, where telecommunications allow the globalization of classrooms. In this sense, the interactive activities that are proposed should involve the application of ICT as a means of construction that allows to extend the mind, to use technologies to learn with them and not from them, and a planned use so that their use is effective and meaningful.

According to Riveros and Mendoza (2005) the principles that guide the relevant use of technology are related to learning in a constructive way and collective knowledge, active, social and individual learning, learning to understand, learning as a social, collaborative, cooperative and socially comparative process; learning as social interaction, socially distributed, situated, generalized and self-regulated learning.

In this process of learning through ICT, students must acquire some skills with respect to digital literacy and are classified into levels. Basic level is where the student must have a general knowledge (written documents, spreadsheets, etc.), know the hardware elements, inputs and outputs of the computer. An intermediate level in which they must be able to create multimedia documents, use ICT as an aid and solve problems. At the last level, ICT are already a tool for the curricular content where the student uses them as a tool to build knowledge and thus obtain significant learning.

Bartolome et al. (2015) argue that the use of ICT by those who transmit knowledge can be in two ways, mandatory or voluntary to expand classes as a resource or tool. When the use is voluntary, the teacher can pretend to have a light workload and be able to change the structure of the subject in terms of content; if the use is mandatory, it can be counterproductive because the teacher is not familiar with the use of ICT. Most of today's teachers did not have technological resources in their educational biography, so the first condition is collaborative work among teachers socializing the results acquired by implementing ICT.

Regarding the teacher who implements ICT, he/she is often pressured by the fact of being innovative and developing flexible proposals that give rise to articulate various curricular spaces, which takes time and in many cases educators must study those spaces that are not in their area; a transmission of values is required to counteract social problems and finally many times teachers are victims of the information society because they are negatively affected by policies that do not contribute in terms of material resources and in addition they have to train and work at the same time causing work fatigue (Bartolome et al.,2015).

As mentioned above, it is important to include proposals mediated by digital technologies in curriculum design and development. The curriculum, when considered as an educational project, is a political-educational proposal that must be developed and interpreted according to the context. It must respect the different cultures, customs and address the realities to which they aspire and have, i.e. its purpose is to improve the quality of life of students.

The idea of taking social problems into account, makes the commitment to social improvement, giving rise to values and not to competition.

Analyzing Herrera and Didriksson (1999), some issues that should be taken into account when designing and developing a curriculum:

- An integrated curriculum should be proposed, based on students' projects and needs, where the teacher's challenge is to make students active, participative and encourage critical opinions.

-A curriculum that allows the elaboration and production of knowledge and cognitive processes by students through learning processes that are adapted to the ways in which our students work (promoting the integration of ICT, for example, since they are digital natives and the working world requires technology).

-Curriculum that allows for flexibility, student initiative, a future time orientation according to social needs.

We understand that ICTs are installed at great speed and are necessary today, causing a great change in pedagogical models, making teachers change methods and resources to improve teaching-learning processes. Technology allows shortening distance, maintaining communication without physical contact, creating a great variety of organizational, technological and pedagogical models to generate teaching based on ubiquitous, instantaneous and sustained communication over time, seeking educational quality with the help of technology.

4.5 Educational Technology

We can speak of a pedagogical discipline called Educational Technology, which emerged in the 50's in the United States due to the confluence of three factors: the diffusion and social impact of radio, cinema, TV and press; the development of studies and knowledge about human learning under the parameters of behavioral psychology and the methods and processes of industrial production. Where the object of study is the introduction of communication resources to make the teaching-learning processes more effective. In the 70's it spreads to other countries, reaching Argentina and is related to the needs derived from industrial technological processes.

At present, educational technologies are being reformulated thanks to the emergence of new paradigms and the revolution implied by the new information and communication technologies; and some ideas about this discipline are as follows: Educational Technology is a space of pedagogical knowledge about media, culture and education in which the contributions of different disciplines are crossed; it studies the processes of teaching and transmission of culture mediated technologically in different educational contexts and the nature of the knowledge of Educational Technology is not neutral nor is it kept aside with respect to the interests and values that exist in the context in which they are implemented.

This technology assumes that information and communication media and technologies are cultural objects or tools that individuals and social groups reinterpret and use according to their own interests. The methods of study and research of Educational Technology are eclectic, combining quantitative and qualitative approaches according to the objectives and nature of the reality studied. Nowadays, the field of study of Educational Technology is the relationships and interactions between Information and Communication Technologies and Education (Area Moreira, 2009).

Another conception of Educational Technology is that it is especially dedicated to the use of audiovisual resources in the teaching-learning process. For this reason, technology is considered a teaching medium that is configured by: physical or material support, the information it possesses, the way in which the information is represented and the purpose it has in education.

Education can achieve its most transcendental goals through the systematic use of educational technology, which employs various means and resources for school learning, whether traditional (books, blackboard, among others), or the tools offered by ICT. It should be noted that Educational Technology is not the same as information and communication technologies, the latter are the digital tools that allow storing, representing and transmitting information and educational technology is the pedagogical discipline responsible for conceiving, applying and systematically assessing the teaching and learning processes, using various means for education to achieve its purposes.

Area Moreira (2009), quoted by Torres Cañizález and Cobo Beltrán (2017) points out that educational technology is a field of study that deals with the approach of all instructional and audiovisual resources; for such reason, the number of technological tools has multiplied exponentially (digital learning activities, portfolios, blogging, among others), designed to energize school environments and promote the acquisition of new skills. Therefore, it is possible to differentiate, since Information and Communication Technologies only group those resources related to the media (cinema, television, radio, Internet) that serve and are responsible for transmitting content with educational value to a group of participants or a society.

Implementing the use of technology in education does not imply increasing its use just for the sake of it, but to be able to detect which are the benefits that technological alternatives could bring to get students to learn better, where the success of ICTs is achieved when teachers incorporate them into the didactic environment, promoting the active participation of students and based on the principles of globalization, interdisciplinarity and transdisciplinarity, using actions that are derived from experiential learning, discovery, projects and problems.

The use of technology should be considered as a goal of today's education and in this way the educational system responds to the needs of the information society, where the efficient, responsible and ethical use of technology, which is essential for survival, either as a citizen or as a worker, in the knowledge society, should be contemplated within the range of skills that students should acquire during their school life.

Everything new implies a challenge for teachers, since technologies are generating structural changes in the organization and development of disciplines, through the incorporation of new learning models.

The use of technology is not only a technical issue; it must raise basic questions and demand new learning models, which must clearly state the educational objectives and, in turn, take into account the new opportunities presented by technologies. The use of ICTs can be used to support traditional learning (as mentioned in previous paragraphs) or to promote distributed learning, which is carried out in both face-to-face and virtual classes, where all the potentialities of technologies are used. Where students not only interact with them, but also through them, and they do so with teachers and classmates where interaction is extremely necessary for certain disciplines.

Using these new technologies requires strategic planning, because at present, although some institutions have an IT infrastructure plan, which is necessary, it is not sufficient to implement new pedagogical models in education. It is necessary to develop plans specifying how the new technological resources will be implemented in the educational process. First of all, it is necessary to see whether ICTs will be implemented solely for the sake of their development, or whether they will be implemented according to the needs of the students and the objectives of the teachers. It is then necessary to plan a teaching model that contemplates all forms: face-to-face, virtual, with shifts, distance or bimodal.

Technologies are a means and not an end, and planning for the teacher does not correspond to a particular curricular space; all disciplines can be taken into account, where in many opportunities it will be more important how it is taught than what is taught. The strategic plan to implement the new educational models based on ICT will be successful if teachers and students are convinced and also have the necessary support for their use. It is also necessary to specify how technology will be used in the teaching-learning process, since the models to be implemented must have the following features

The support of the institution and the entire teaching team must work together to achieve the proposed goals. It is very important to take into account the monitoring of the proposed model in order to detect failures and thus be able to restructure based on the needs on an ongoing basis.

The use of technologies is being imposed on teaching models, and a structural and cultural change is required to uproot habitual practices. Although the implementation of new technologies is at an initial stage, a clear vision of change and attitudes of the current teaching model must be approached from the outset.

In this sense, the ways of teaching are evolving, thanks to the incorporation of ICT and the new paradigms, since students dominate mostly diverse social networks and information, where different types of learning take place. In the year 2020 (pandemic year), multiple advantages of technologies are known and used to organize contents, evaluations and control activities.

Similarly, the properties of the platforms or applications offered to students should allow them to access quality education, and for this to happen, teachers must be equipped with tools that allow them to complement the content of the curriculum design with information and communication technologies, so that each educator can use different programs and configure a virtual classroom that is impactful for their students.

Teachers should teach through ICT, student-centered and with the pertinent curricular adaptations, appropriate virtual platforms, bearing in mind that new technologies must be accompanied by good pedagogical planning, innovative strategies and resources, otherwise they are a waste.

To innovate in educational projects based on ICT, it is necessary to eliminate the digital divides mentioned in previous paragraphs, in which there are or exist several dimensions: economic (since not everyone has access to technology), political (ICT regulation must be taken into account), technological (resources in terms of the network), social (the population that cannot access) and cultural (thoughts and attitudes towards technology).

The use of EVEA technological resources allows students to interact with the teacher and their classmates, depending on the platform used, it can be through comments in a conversation thread, through forums, active videoconferences such as the Zoom application[4] or from the Classroom with Meet[5] . These teaching-learning environments are free of time and space restrictions in face-to-face teaching and can ensure the continuity of virtual communication between teachers and students, complementing face-to-face teaching with virtual activities.

---

[4] Cloud-based videoconferencing service that you can use to meet virtually with others, either by video or audio only or both, allowing for live chats and recording those sessions for later viewing.
[5] Videotelephony service developed by Google

Thus, there is a need for a new educational system, new educational policies, new scenarios and materials, new ways of organizing educational methods that allow an education according to the needs of students based on the current society and technological advances, for which educators trained in network didactics are needed.

These resources are important and enriching if they are used as a means of information because they allow informing and disseminating, promoting the exchange of opinions, visualizations and immediate responses to participants, and as a means of relationship because information can be exchanged between users about different interests Area, M. and Adell, J. (2009).

Through digital resources it is intended that students learn to read and interpret the proposals that both teachers and students offer, respecting and duly and adequately justifying each position, giving priority to reasoning about the different situations that need curricular content to be solved.

In order to engage our students, it is necessary to use attractive digital resources that in turn provide a timeless and flexible communication that allows us to be a positive option in the face of absenteeism that some members of the student body may present for different personal and/or social reasons.

The instruments to be proposed should speed up the practices, therefore the use of devices as pedagogical tools should be encouraged, allowing an organization of the activities, providing the student with an order of the subject and a constant updating of ideas and resources.

In the framework of this great connectivity that develops in a network and generates knowledge in today's society, by what is called "Social Constructivism" allows knowledge to be generated in a dynamic and changing way where the person learns through the internalization of socially constructed knowledge (Gros et al., 2012).

Given the free access to a great variety of data, students must be taught to select the sources of information, giving meaning to what they intend to learn, being the student the one who builds their learning, offering the teacher to be their support and support, that is to say that both teachers and students must work together.

In this sense, "Pedagogical Theories" are proposed to configure new teaching-learning processes mediated by Information and Communication Technologies, to include technology in the curriculum and to focus education purely and exclusively on the student and not on the teacher.

The integration of ICT should be done gradually from the curriculum and didactics (Sanchez, 2002), that is, to make them part of the curriculum with a harmonious and functional use for the purpose of learning specific to a school discipline. The combination of ICT and teaching moves the student to a new place of entertainment (Merril, 1996) cited by Jaime Sanchez Ilabaca (2003), their uses should be promoted to stimulate learning, integrating them as a tool that is linked to the curriculum and with a curricular objective.

In the area of mathematics ICT should allow the acquisition of knowledge, improving the students' ability to solve different situations through different solutions or paths giving great importance to problem solving, using technological tools to interpret, process and solve different issues related to the area, this implies that it can be included in the curriculum as an evaluation criterion the fact of having skills and knowledge of technological tools in a meaningful context Arrieta(2013).

ICT can be included in mathematics by means of images, graphics, use of software, spreadsheets, etc. Allowing students to experiment, conjecture, correct and manipulate information, thus being able to materialize through different simulators concepts that are often abstract.

If it is proposed to teach mathematics using ICT, the teaching-learning process is favored in several aspects such as understanding and improving their learning, observing concepts through software that allows them to interact by manipulating elements with which they can understand properties, organizing and analyzing data, interpreting different units of measurement by visualizing figures and bodies, etc.

It is important for teachers to research and use different resources found on the web to implement and work in the classroom, whether face-to-face or virtual. Managing a class using ICT makes learning more homogeneous with the use of different software specific to the subject, where dynamic resources allow a consolidation of concepts and properties favoring reasoning, also acquiring visual perception, linguistic and digital competence.

ICT can play a very important role in the process of teaching and learning mathematics, but only if they are used correctly. Moreover, if their use is not adequate, they can trace a tortuous path from being a powerful tool to a barrier that impedes the process.

4.7 ICT and Mathematics

Both the teaching and learning processes are different, therefore, sometimes the same ICT cannot be used in both processes. In the teaching process, the group of ICT tools will be composed of specific tools for the subject or for education in general. Thus, the digital whiteboard, as far as hardware is concerned, can be a good ally of the teacher because of its possibilities. As for software or specific applications, we could mention, with an eye on free software, the following: Xmaxima, GeoGebra, Kig, Kmplot, Geomviewe, etc.

Although students are very familiar with technological resources and teachers have been trained, we should not be afraid to implement them in the classroom because the objective is to teach them and have them learn mathematics and not to teach them how to use ICT. This means that we intend to rely on the prior knowledge of our students to achieve the objective(s) set in the classroom. This previous knowledge of the student must be used and know how to lead to achieve the objectives we propose. Mathematics can be learned in an idle way with some of the resources that can be used to contribute to the learning of mathematics with the use of ICT. On the Internet and in specific educational repositories we can find various tools that can help students in the learning process and for which no computer knowledge is needed (Revelo and Carrillo, 2018).

One of the main objectives that we have as teachers in the development of mathematics, is the change of attitude to take steps to initiate problem solving, giving students the ability to perform a proper interpretation of different types of information, where they themselves can find the solution to the problems that arise in their daily lives, this is one of the skills required by the student. In turn, with the amount of information that can be obtained on the web, mathematics, which is our case, has been greatly benefited, because there are a lot of software that allow students to improve the visualization of concepts and ensure an understanding of them.

The most important influences that are allowing the use of ICT in the teaching-learning process is the dynamic interaction of students in the curricular space of mathematics because it allows to build learning focusing on the digital environment where the different software allows to solve proposed problems, improving the ability to investigate and information management.

There is a great variety of web pages, which allow the student to learn and have fun developing the exercises. In these pages you can find topics such as algebraic operations, numbers and operations, geometry, equations, statistics and probability, problem solving. One of the most used softwares in mathematics is GEOGEBRA, of free access, highly recommended for the teaching of mathematics in educational institutions, within its work areas are geometry, arithmetic, algebra, calculus, functions, statistics, trigonometry and probability. It has the particularity of having a very large database of activities, simulations, exercises and lessons. It is a very useful tool where students can expand their knowledge, discovering new ways to interact and relate their basic mathematical concepts. It is important to emphasize that ICT serves as a support for those students with cognitive difficulties so that when they join a group they can support and complement each other; to find the most appropriate way to perform the activity that the teacher asks them to do.

The uses of web browsers allow locating, filtering, organizing, evaluating and classifying information and content, allowing the development of collaborative learning in the area of Mathematics, promoting interaction through the management, use and application of digital communication (Revelo and Carrillo, 2018).

Digital environments allow creating and managing specific contents of the mentioned curricular space, also specific free software allow making modifications according to the needs of the students in the teaching-learning process of mathematics. ICT allow building mathematical ideas and concepts facilitating learning by discovery (Revelo and Carrillo, 2018).

Thus, it is important to inquire about the characteristics of the different digital tools available and, based on this, to foresee activities with ICT that allow the student to have a constructivist role, promoting individual, collective and/or collaborative work, encouraging debate from different points of view, promoting continuous feedback both in the practices and in the evaluation process. At the same time, they should be provided with cognitive tools and competencies that allow them to act in a critical, creative, reflective and responsible way on the abundance of data, so that they can apply them to different contexts and learning environments and thus build knowledge, adapting the use of digital resources to obtain different didactic and methodological strategies.

4.8 Evaluation

The way to account for the student's acquisition of skills and knowledge is by means of evaluation. Evaluation implies a judgment on the quality of learning obtained by the student as a consequence of participation in different teaching-learning activities. This means that from a certain proposed activity, the idea is that the student incorporates new knowledge by working together with the teacher.

There are different evaluation practices, the first is the initial one, i.e. the diagnostic evaluation, which is carried out at the beginning to collect previous knowledge and thus establish where to start from. Formative evaluation, which is carried out during the teaching-learning process and allows us to improve the teaching performance, is regulatory since it allows us to adjust the educational process according to the needs of the students. In this type of evaluation, the student must be aware of the process of knowledge construction and the object of learning, and the final formative evaluation serves to indicate whether or not the student finally achieved the minimum objectives of the space. Evaluative practices are composed of different levels: the program, the activity and the evaluative task.

In this context, we must consider what deserves to be evaluated, how and for what purpose, projecting an evaluation program that allows us to account for the different ways of collecting information on the learning process of the students, the frequency and order of these activities, the time in which they are placed within the teaching process and the use given by the students and the teacher to the information collected in these evaluations.

There are different stages within the evaluation that we must take into account, for example: the activities that are prepared for the students to participate, the activities that are found in the evaluation, the corrections, the communication and the evaluation of the different answers to generate new tasks. The concept of evaluation is very broad, it can be defined as a collection of information by the teacher that allows to give a value judgment on the learning that students achieve as a result of the participation of the proposed teaching activities.

Iturrioz and González (2015) argue that on several occasions what is taught is confused with what is learned and this is reflected when a teacher says that he/she is going to evaluate what he/she taught, without recognizing that the student transforms or reconfigures what was transmitted to him/her.

The new technologies provide the possibility of an evaluation based on transparency in the debate, exchange and discussion of the participants, bringing three important changes. These are:

- automatic evaluation, in which the students automatically obtain the answers and corrections, such as tests
- encyclopedic evaluation when they must prepare research and

development work where there is easy access to information sources.

- collaborative evaluation where the process is evaluated as well as the product, promoting shared responsibility.

The contribution of ICT in the teaching-learning process is the different potentials to develop such process. In addition to the great connectivity they provide, there is a formalism since there are different steps to follow, precise instructions to access, process and transmit, with respect to this they favor cognitive and metacognitive processes related to planning and executing actions according to an established plan based on technological requirements.

There is interactivity allowing face-to-face interaction according to needs, allowing the student to build his position in the face of different opinions, thus improving self-esteem when he manages to build a meaningful learning. There is dynamism because it allows representing phenomena through simulators that lead to a better understanding. Multimedia resources are generated because it is possible to create environments that combine different information representation formats. Another characteristic that the new technologies allow is the incorporation of hypermedia, since the student can explore the information autonomously and according to his interests.

According to Lafuente Martínez, M (2003), designing evaluations by means of technological tools leads to thinking about pedagogical design and instructional activities from a pedagogical and technological point of view, which is the origin of Technopedagogical Design.

First of all, it is necessary to approach the study of the different devices and their properties used in the moments of evaluation practices in the real use given by teachers and students around the contents, tasks that are addressed, since the pedagogical and technological interaction is a complex relationship and a coherence between both is not ensured. It is not only based on the technological resources available, but also on the planning of activities and the effective pedagogical use of these resources by students.

Consider the criteria to be taken into account on which devices are more or less concrete, the design of coherent activities and which are integrated taking into account the objectives of the subject, which is retroactive allowing to give rise to new discoveries of contents and learning, allowing to develop personal and autonomous environments. Activities that gradually allow knowing what the student has built as knowledge by getting involved in the proposed activities and not only considering the evaluation for grading.

5. PROPOSAL

The diagnosis carried out allows us to affirm that sometimes the proposals made by teachers with technological resources in the classroom are not very attractive for students. Therefore, it is proposed to implement a virtual classroom for the development of the Mathematics Curricular Space, easily accessible and free of charge, allowing students to have access to knowledge in an organized manner, to have access to class material at any time and place, to consult different interactive resources and to make consultations on all the proposed knowledge.

In our case we will choose to work with Classroom, this resource is an educational social network that allows a dynamic approach and easy navigation, among the panel of activities it has, you can organize classes, topics and assign tasks, roles, organize groups of students and has the advantage that all information and / or material is automatically saved in the Drive.

It allows working collaboratively with shared documents and establishing links between teachers and students, favoring debate and the construction of knowledge.

5.1 Organization

Based on the proposal and the objectives previously stated, an EVEA is developed and proposed in which students can access the knowledge corresponding to the area and year of course, conceptual schemes (figure 1) on the subject as a whole as well as on each unit to be developed (real numbers, linear function, systems of equations and trigonometry). The course provides spaces where students can consult links to different resources, activities that contain the basic knowledge proposed by the NAPs, interactive proposals that allow them to detect errors in order to restructure the following classes taking this as a source of knowledge, activities to be delivered, etc. The access code to register for the 2021 course is: **mnjm6lz**, and for the current year (2023): **5scqbjjj.**

The selected platform (Classroom) allows assigning roles, there can be more than one teacher, which made it possible to incorporate the

Inclusion Support Teachers (DAI) (Figure 2), for the adapted work to be assigned to their students, allowing them to work autonomously and the students to be comfortable with individual and personalized work assignments.

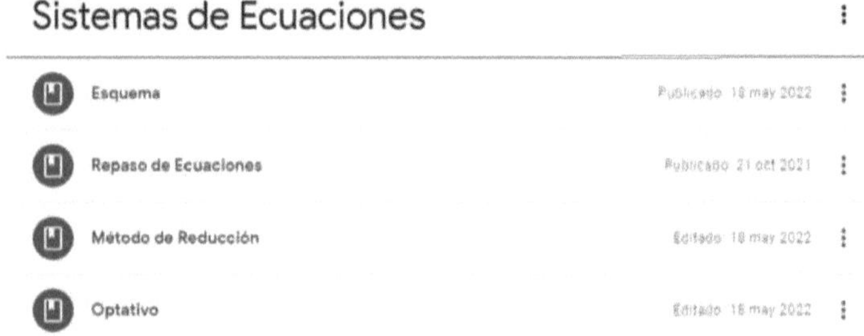

Figura 1: Desarrollo del tema Sistemas de Ecuaciones y su esquema conceptual.

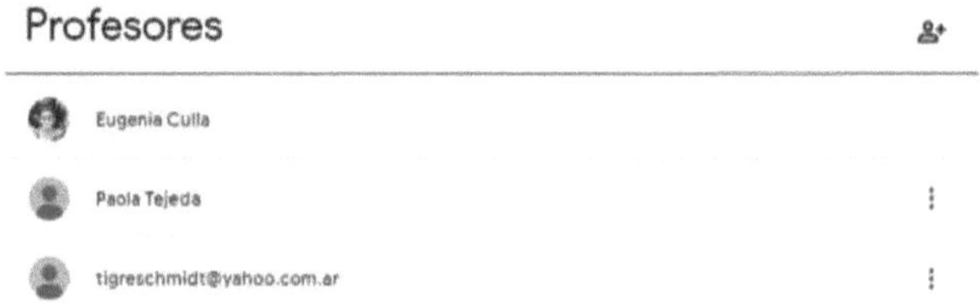

Figure 2: Classroom teaching team consisting of the teacher, teaching assistant and IAD.

Classroom has a very accessible design and operation, it has tabs organized in news (main wall where notifications appear), class work (shows the list of topics and what each one contains), people (class members) and grades (student performance, a tab that is not in the interface that appears to students) although it does not have evaluation tools, the Google Form can be used or resources can be incorporated as materials that allow interactive exercises. In this resource it is possible to create from tasks to materials as shown below (figure 3).

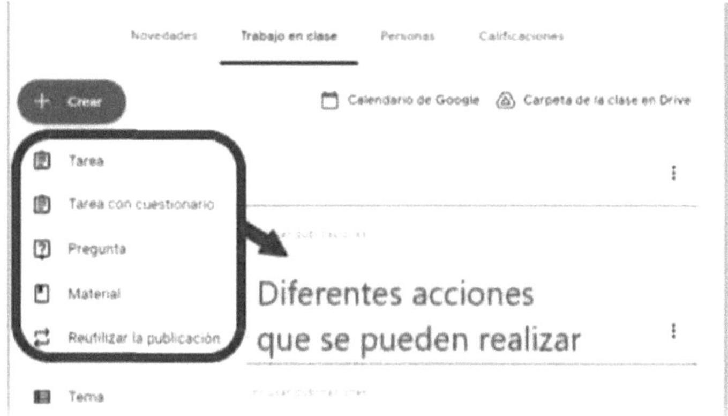

Figure 3: List of actions that can be proposed from the teaching role.

In the material section you can offer resources in different formats (PDF, conceptual charts), interactive review activities, practical work to be developed. In the same way, in the homework section, different activities can be presented, but with the purpose of having a formality of obligatory delivery.

As the use of different Google applications requires students to have a certain email (in this case Gmail), the first step in the classes of the subject will be to create workspaces for digital literacy on the creation and management of email accounts, adaptation of the different work devices (computers, cell phones, tablets, etc.). Also, to go through the EVEA proposed by the curricular space, knowing the different ways of registering in it (by invitation or with a code) and the variety of resources it has, as well as the use of applications and/or programs relevant to the curricular space.

Likewise, each proposed activity is organized by themes, allowing for a good organization of the classroom (Figure 4).

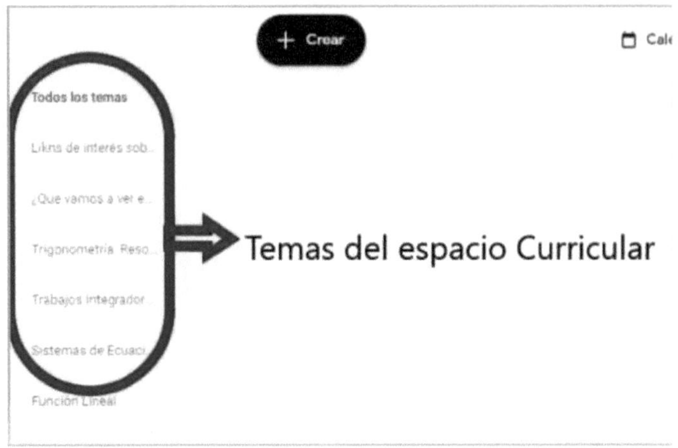

Figure 4: Classroom organization by curricular space topics

In terms of organization, the students adapted to the platform, used the possibility of making private comments on the proposed activities as well as comments on the communications of news and queries in the places where the materials were published (Figure 5).

Figure 5: Submission of online activities and comments on the proposed activity

5.2 Synchronous Classes

In the proposed virtual space, the platform is organized in topics, one of them refers to the live virtual class link: "Link for Virtual Classes" (action that was carried out in 2020 and the months from May to July 2021 when two policy measures were implemented: ASPO[6] and DISPO[7]), the resource to be used in this case is ZOOM. This platform allows videos, video conferences, where they can be group or individual, allows screen sharing, allowing both teachers and students to show and interact in the correction of exercises and / or proposed situations, can be recorded, allowing a richer resource with the interaction of stakeholders allowing students who for various reasons can not join synchronously access the material at another time. It also has chat and the possibility of sharing files at the time of the meeting. As a security feature, there is a waiting room, where the teacher allows only those students who correspond to the class to enter (Figure 6).

Figure 6: Link and schedule of virtual classes

In this sense, the link generated for the online classes is recurrent, thus allowing to reuse it if the forty minutes allocated at that time are not enough, they are recorded and then uploaded under the title "CLASS OF (Date)" (Figure 7), remaining as a resource where students who were not present at the meeting can obtain the material, and those who participated

---

[6] Social Isolation, Preventive and Mandatory.
[7] Distancing, Social Preventive and Mandatory.

can review it in case they need it for self-correction of the proposed activities.

Figure 7: Files of the synchronous virtual classes.

Since the information remains in the EVEA, it can be used every time the student requires it, promoting feedback and providing a model of an inverted class if we consider that the students can actively participate asynchronously, note down and list those doubts that arise for consultation in the next synchronous class.

With respect to the synchronous classes, we can say that out of a total of 23 students, between 7 and 13 attended the classes that were given once a week according to the organization of the schedule presented by the school (Annex II).

5.3 Proposed Activities

Among the activities proposed is the resolution of different practical tasks that allow the acquisition of mathematical knowledge by using the platform and different resources, with the objective that they can manipulate the different tools to be active participants by consulting, debating, observing and carrying out the proposed activities (Figure 8).

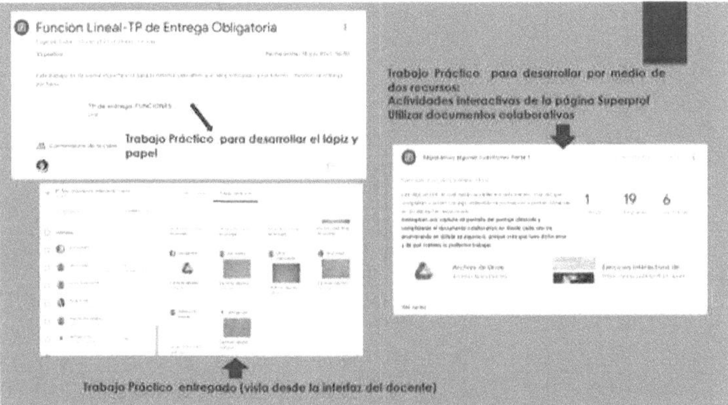

Figure 8: Proposed activities using ICT resources

Present a list of different technological resources to cover the different topics proposed in the NAPs (Figure 9), as well as a schedule of use and availability of the resources the educational institution has to implement ICTs both in the classroom and outside it (mobile classrooms, projector, etc.) and thus achieve an institutional organization for the use and care of technological devices.

Figure 9: ICT resources to carry out different activities related to the subject.

The elements of the list of knowledge will be completed as the classes take place, whether face-to-face, virtual or combined, since within each proposed and ordered knowledge will be found:

- Weekly classes: videos where teachers and students interact in the explanation, development and correction of activities proposed by Zoom. This resource allows students who connect through mobile devices to interact in written form through the touch screen, or show with their camera the situation in which they have doubts, allowing the teacher to take a screenshot, and show it on Paint, a resource that allows editing the photo making the necessary corrections. They are uploaded to EVEA under the material label.

- Practical Work: activities to learn the proposed knowledge, with practical activities produced by the teacher.

- Interactive activities: by means of pages with mathematical contents, whose delivery is compulsory, as well as the elaboration of collaborative documents: reports that allow the revaluation of errors as a source of learning (Figure 10).

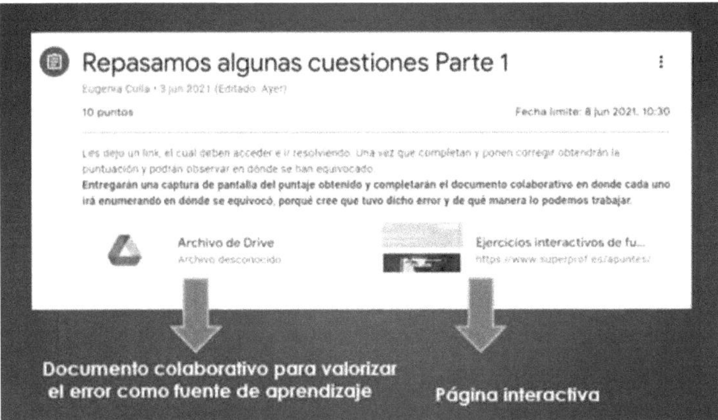

Figure 10: Activities with interactive pages and collaborative documents.

- Compilation of resources that students use as support in addition to those proposed by the teacher, such as YOUTUBE videos (Classroom code: **5scqbjjj** resources requested by a group of students), as they are easily accessible and often provide a rich discussion when someone finds a different method proposed by the teacher to solve the activities given in class.

- Provide a conceptual outline of the curricular space so that students can organize the information provided (Figure 11), and the possibility of having them create such conceptual outlines by topic or trimester.

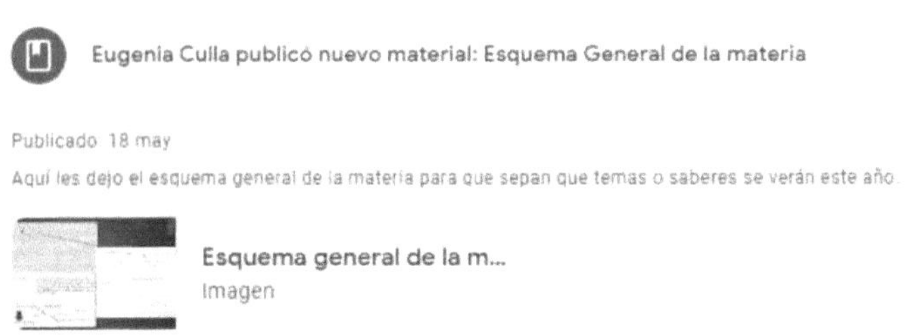

Figure 11: Scheme of knowledge to be developed during the school year.

- Establish times and physical or virtual spaces so that students can give an account of the information they possess and the knowledge they can acquire, fostering self-confidence and combating the prejudice that mistakes are frowned upon, but rather promote them as a source of knowledge and a starting point to achieve constructive and meaningful learning (as shown in Figure 9).

- Promote the search for interactive pages of interest to students so that they can incorporate them into the platform offered to them, making them participate in the design of the different tools proposed for the classes and also encourage collaborative work in the preparation of reports.

- Provide online resources such as virtual homework, videos and/or conferences for extraordinary cases such as flexible trajectories (a situation that occurs when students do not attend school physically for a long period of time), providing pedagogical assistance not only to the student, but also to the home teacher appointed by the ministry to provide support.

With respect to what was stated in the previous paragraph, in EVEA several resources are proposed that are incorporated as delivery work and that allow interactive activities to be carried out, giving the corresponding corrections at the same time. One of them is "Superpfrof", which corresponds to a project that was born in France and spread almost all over the world, reaching Argentina as well. Although it is a paid online tutoring service, in the Spanish domain there is an interface https://www.superprof.es/apuntes/escolar/matematicas/ that provides theoretical summaries, explanations, activities to solve and interactive activities on the different subjects taught at the intermediate level, it has the option to correct, where it provides instant corrections allowing the student to account for the mistakes made (Table 1).

Analytical Geometry	Conics-Distances-Rects-Vectors
Arithmetic	Integer-Natural-Proportionality-Rational Real - Decimal Metric System Successions Complex numbers-Decimals-Divisibility
Trigonometry	Trigonometry
Algebra	Polynomials-Equations-Equations-Logarithms
Linear Algebra	Systems of Equations-Programming Linear Determinants - Matrices

Table 1 : Topics provided by the page regarding different areas of mathematics.

- Activities are also proposed with "Quizizz"[8] , which are questionnaires that can be carried out synchronously or asynchronously, providing immediate feedback to the students' answers, do not require installation and are free of charge, and teachers can create their own activities or use a repository in which they can also modify the existing ones according to the needs of the group (Figure 12).

---

[8] https://quizizz.com/

Figure 12: Integrating activities of operations with decimal expressions and linear functions proposed in Quizziz.

With respect to the aforementioned, review activities were proposed (figures 13-1415) not only to observe the knowledge obtained and reinforce those unfinished, but also to learn about the proposed platform and then a closing activity for which some activities had to be solved on paper to then dump the result in the ICT resource, where the results obtained are observed (figures 16) allowing to review what knowledge or concepts each of the students have or have not acquired.

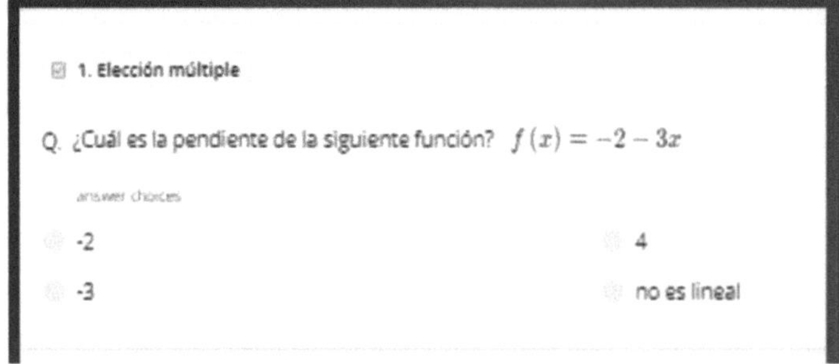

Figure 13: Activity 1 to identify line parameters

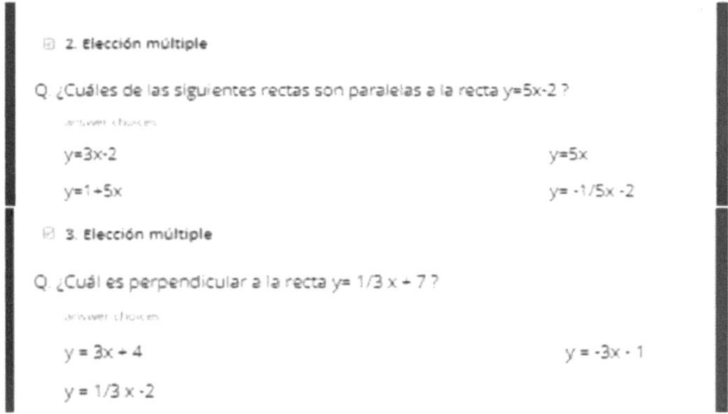

Figure 14: Activities 2 and 3 on the concept of parallelism and perpendicularity.

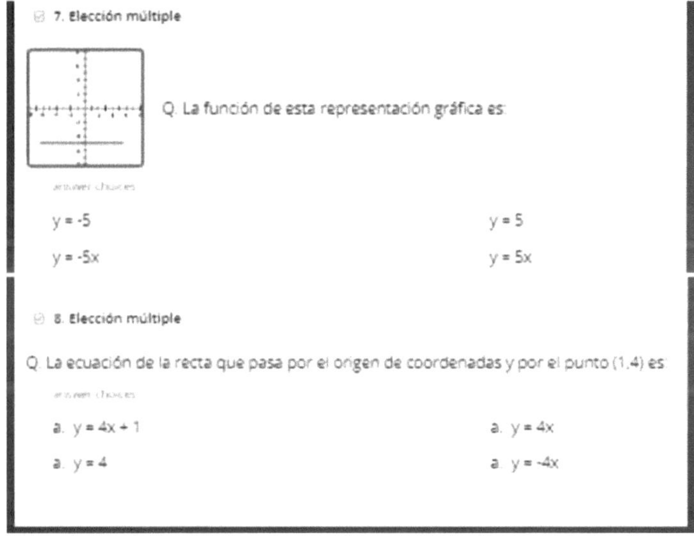

Figure 15: Activities 7 and 8 on graphical representation and line passing through two points.

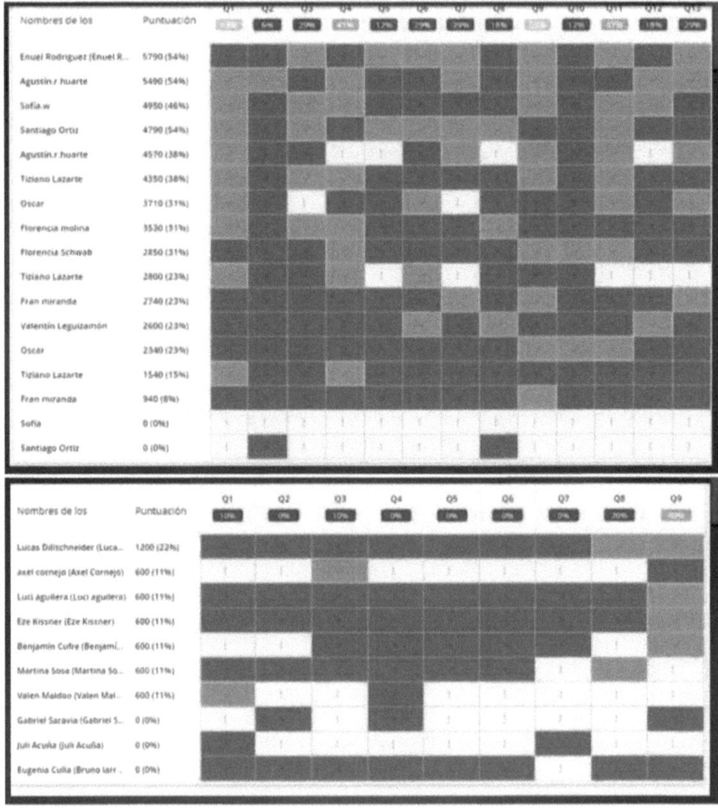

Figure 16: Detail of correct and incorrect answers for each student.

Other interactive tasks that can be proposed is through the resource That Quiz[9] , which is a 2.0 tool that offers a free randomized quiz service that gives the results of the quizzes immediately. It allows you to upload the list of students by creating a class so that the scores are automatically imported. The activities can either be created or use those created by other teachers using the repository (Figure 17).

---

[9] https://www.thatquiz.Org/es/#

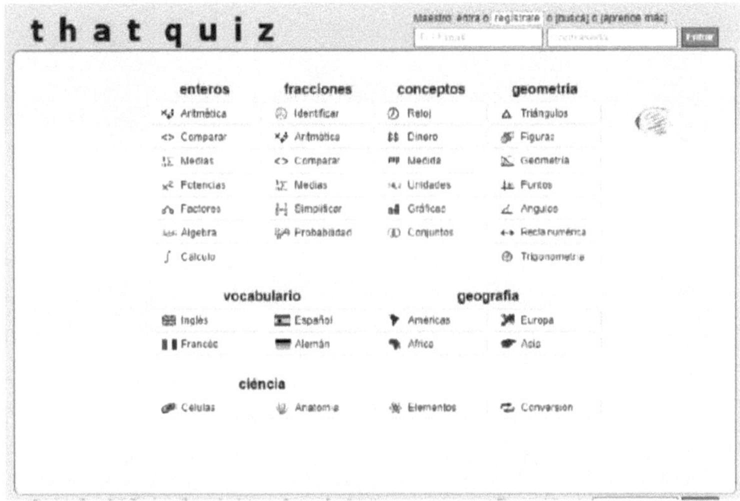

Figure 17: ThatQuiz interface

These three proposed tools allow us to account for the students' process in the acquisition of knowledge, as well as to implement the evaluation as a process, since, interspersing the explanatory classes, correction and development of the proposed activities with games involving exercises or questions that account for the acquisition of knowledge, concepts and strategies for solving problematic situations.

As a way of organization, as we are in a virtual way, the meetings with the students are once a week, on the platform the day they do not have virtual class and respecting the schedule of the subject, they will be reminded on the news wall of the activities we are doing and the virtual meeting of the next class. In this way we have a kind of organization and routine with respect to what is a totally face-to-face schooling and interact through asynchronous consultations without losing the pedagogical link.

5.2 On some of the results obtained:

The general objective of designing and implementing a virtual classroom in the area of Mathematics was achieved, although students adapt to the implementation of new technologies in the classroom by using the proposed platforms, there are aspects that need to be deepened, such as the conscious use of technological resources as a study tool, the use of the platform outside school hours, if necessary, as a means to resolve concerns.

In the 2021 academic year, audiovisual material from YOUTUBE was not used since the synchronous virtual classes were recorded and uploaded to Classroom. In the 2022 academic year, some students requested material to deepen their knowledge during the diagnostic stage in March and were provided with audiovisual material and review activities: https://www.youtube.com/watch?v=PnATAsxu_oo (Figure 18). It was proposed to work with interactive pages that contemplate the contents included in the Priority Learning Nuclei, where students engaged with the proposed activity online. However, better results in participation are obtained if the same task is carried out through technological resources, but in a face-to-face manner, i.e. in the classroom.

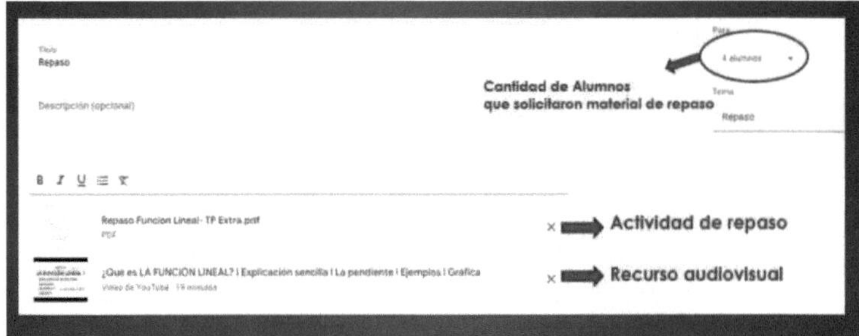

Figure 18: Activities requested by students.

The aforementioned activities were modified according to the classroom proposals, changing some of the instructions already established in the repository of the platform used and others were created according to the needs of the students.

With regard to the virtual classes, after arduous work to raise awareness of the importance of participating in them, a considerable number of students were able to participate in them, ask questions and, on occasion, turn on their cameras to show their concerns expressed on paper based on the proposed activities.

In the year 2021, bimodality was implemented for a time (classes alternating between face-to-face and virtuality), which led at that time to change certain strategies and take advantage of the face-to-face space for the use of ICT, since it is important that the fact of being present in the classroom should not prevent the implementation of virtual spaces to be promoted so that students achieve habits of participation through different technological resources. Thus acquiring another way of learning outside the traditional, not only in the classroom with a debate and consultations, but exposing the activities that are proposed losing the fear of being wrong and understanding that in learning communication in all media is important to acquire knowledge.

Based on the proposals made, the participation of students in collaborative proposals was achieved (Figure 19); in the interactive activities proposed with the Quizziz resource, which were more attractive to them than those given through ThatQuiz and Superfrof.

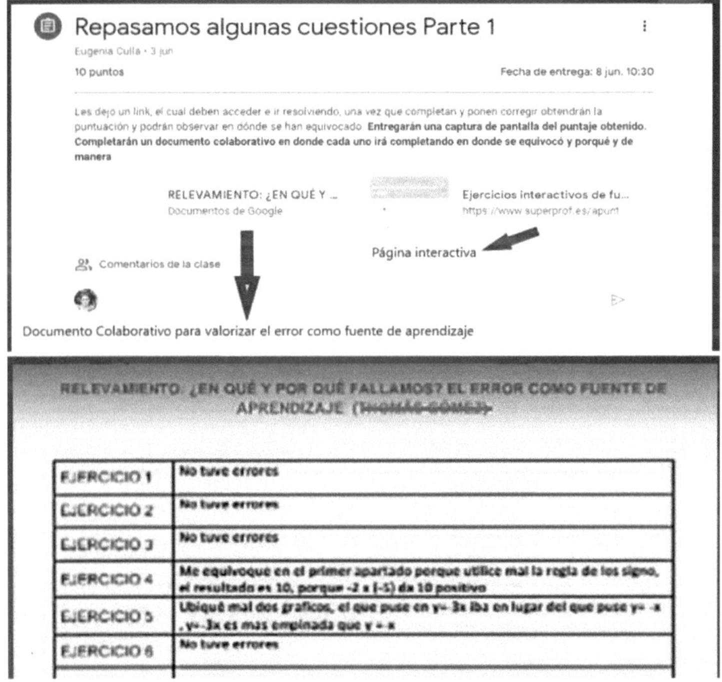

EJERCICIO 1	No tuve errores
EJERCICIO 2	No tuve errores
EJERCICIO 3	No tuve errores
EJERCICIO 4	Me equivoqué en el primer apartado porque utilicé mal la regla de los signo, el resultado es 10, porque -2 x (-5) da 10 positivo
EJERCICIO 5	Ubiqué mal dos graficos, el que puse en y= 3x iba en lugar del que puse y= -x , y= 3x es mas empinada que y = x
EJERCICIO 6	No tuve errores

Figure 19: Proposed activity and screenshot of the collaborative document responses.

In the case of Classroom, it allows the observation of the deliveries through the classification chart (figure 20); on the other hand, through the digital resource Quizizz, it is possible to observe in detail what needs to be reinforced in each student in particular, obtaining a more specific follow-up regarding the acquisition of knowledge (figure 21).

Ordenar por apellido ▾	23 ago 2021 Trabajo Integrad… de 10	16 ago 2021 Trabajo Integrad… de 10	15 jun 2021 Función Lineal-T… de 10	8 jun 2021 Repasamos alguna… de 10	15 abr 2021 Repaso de 10
Promedio de la clase			6.83	8	6.67
lautoro 200	Sin asignar	Sin asignar	Sin entregar	Sin entregar	Sin entregar
thomas games 2005	Sin asignar	Sin asignar	Sin entregar	Sin entregar	Sin entregar
Juli Acuña	Sin asignar	Sin entregar	Sin entregar	Sin entregar	5
Luci aguilera	Sin asignar	Sin entregar	7	6	5
Danilo Balduini	Sin asignar	Sin asignar	Sin entregar	Sin entregar	Sin entregar
axel cornejo	Sin asignar	Sin entregar	9 Entrega tardía	Sin entregar	5 Entrega tardía
Alicia Isabel Demasi	Sin asignar	Sin asignar	Sin entregar	Sin entregar	Sin entregar
Lucas Dillschneider	Sin asignar	Sin entregar	7	10	Sin entregar
Florencia Echev	Sin asignar	Sin asignar	Sin entregar	Sin entregar	Sin entregar
Maria Ines Fassi	Sin asignar	Sin asignar	Sin entregar	Sin entregar	Sin entregar

Figure 20: Trajectory of student delivery.

Figure 21: Individual student responses.

According to Rodríguez, D. (2020) ICTs in everyday life are of utmost importance since they allow breaking physical barriers of communication allowing connecting people from different parts of the world with similar interests, creating spaces for cultural, educational and political exchange, etc.; they provide updated information in real time and allow performing different types of activities that make the life of any citizen (paperwork and administrative procedures) in labor, health, education and business.

Although the aforementioned is a work that takes time, in order to be concrete and successful, it must be considered as an institutional proposal, that is, a collaborative work from the teaching team.

As mentioned above, the objective of all disciplines is for students to acquire technological skills that will allow them not only to acquire the knowledge proposed by the school through each subject, but also to acquire the knowledge and skills that will allow them to develop the necessary knowledge and skills to be able to work in a more efficient and effective way.

The goal of the program is not only to provide them with a curriculum space, but also to allow them to apply their knowledge and skills in everyday life, either to enter the labor market, where major human resources agencies nowadays conduct virtual interviews and enrollments through platforms, or to continue studying.

6. REFERENCES

Adell, J (May 13, 2018). ICT breaks the walls of the school. Teacher's Web. CMF. https://webdelmaestrocmf.com/portal/jordi-adell-las-tic-break-the-walls-the-school/.

Alcántara Trapero, D. (2009). Importance of ICT in Education. Innovation and Educational Experiences. Number 15. Sevilla.

Area, M. and Adell, J. (2009): -eLearning: Teaching and learning in virtual spaces. In J. De Pablos (Coord): Tecnología Educativa. La formación del profesorado en la era de Internet. Aljibe, Málaga, pp. 391-424.

Area Moreira, M. (2009). Introduction to Educational Technology. University of La Laguna Spain (pp 15-23).

Arriaga, Elias J. (2003). Las TIC y las matemáticas, avanzando hacia el futuro [Master's Thesis, University of Cantabria]. https://repositorio.unican.es/xmlui/handle/10902/3012

Arrieta, J. (2013). ICT and mathematics moving into the future. University of Cantabria. Faculty of education. España.

Bartolome, V.; Caram, C.; Los Santos, G.; Negreira, E.; Pusineri, M.(2015) Writings in the Faculty: Pedagogical Reflection. Edition III. Ensayos de Estudiantes d ela Facultad de Diseño y Comunicación. Vol. 109. Year 11.

Burin, D., Coccimiglio, Y., González, F., & Bulla, J. (2016). Recent Developments on Digital Skills and Reading Comprehension in Digital Environments, 6(1), 191- 206.
Retrieved from http://revista.psico.edu.uy/index.php/revpsicologia

Camero Almenara, J (2014). Reflections on the Digital divide and education. University of Seville (Spain -UE)

Castells, M. (2009) Comunicación y poder -Alianza Editorial-Madrid ISBN, 97884-206-8499-4

Cachay Osorio, M. (2019). "Importance of the implementation of ICT in educational institutions in the teaching of mathematics". Lima Peru.

Cobo, Cristóbal (2016). La Innovación Pendiente, Reflexiones (y

Provocaciones) sobre educación, tecnología, y conocimiento. Ceibal Foundation Collection/Debate: Montevideo.

Crespo, M., & Palaguachi, M. (2020). Education with Technology in a Pandemic: Brief Analysis. Revista Scientific, 5(17), 292-310, e-ISSN: 2542-29.

Díaz, Rubén (2009). "What if education happens anytime, anywhere?", Expanded Education (pp.51-66). http ://www.zemos9 8.org/eduex/spip.php?article 171.

Fernández Aprile, L. (2020). Higher Education and Technology: Historical evolution in Argentina and the social context in times of Pandemic. HOLOGRAMATICA. Year XVII Number 32, V1. pp. 163-180.

Fernández Tilve, Mª Dolores, Álvarez Núñez, Quintín, Mariño Fernández, Raquel Elearning: Another way of teaching and learning in a traditionally face-to-face University. A particular case study. Profesorado. Journal of Curriculum and Teacher Education [online]. 2013, 17(3), 273-291[date Accessed July 6, 2021]. ISSN: 1138-414X. Available at: https://www.redalyc.org/articulo.oa?id=56729527016

Flores Romero, M., Aguilar Barreto, A., Hernandez Peña, Y., Salazar Torres, J., Pinillos Villamizar, J., Pérez Fuentes, C. (2017). Knowledge Society, Tics and its influence on Education. Revista Espacios (Vol.38. Number 35 p. 39).

Forero de Moreno, Isabel THE SOCIETY OF KNOWLEDGE. Revista Científica General José María Córdova [online]. 2009, 5(7), 40-44 [date of consultation April 26, 2021]. ISSN: 1900-6586. Available at: https://www.redalyc.org/articulo.oa?id=476248849007

Gomez, J. (2020). Google Classroom: A tool for pedagogical management. Mamakuna Revista de divulgación de experiencias pedagógicas. pp45-54. International University of La Rioja.

Granados-Romero J, López-Fernández R, Avello-Martínez R, Luna-Álvarez D, Luna-Álvarez E, Luna-Álvarez W. (2014).

Information and communication technologies, learning and knowledge technologies, and technologies for empowerment and participation as tools to support the 21st century university teacher.

Medisur, 12(1): [approx. 5 p.].

http://www.medisur.sld.cu/index.php/medisur/article/view/2751

Gros, B. et al. (2012). Knowledge Society. Pedagogical Perspective. Chapter 1 of Lorenzo Aretio's book "Knowledge Society and Education".

Gutiérrez, L. (2012). Connectivism as a learning theory: concepts, ideas, and possible limitations, Revista Educación y Tecnologías, 1, pp. 111-122.

https://dialnet.unirioja.es/descarga/articulo/4169414.pdf

Herrera, M.; Didriksson, A. (1999). La Construcción Curricular: Innovación, Flexibilidad y competencias. Higher Education and Society. Vol. 10(2), pp. 29-52.

Ilabaca, J; Curriculum Integration of ICTs Concept and Models. Enfoques Educacionales Journal 5 (1): 01 - 15, 2003.

https://enfoqueseducacionales.uchile.cl/index.php/REE/article/download/4751 2/49550/ 168470

Iturrioz,G; Gonzalez, I. (2015). Evaluar en la virtualidad. pp.133-144.

https://p3.usal.edu.ar/index.php/signos/article/view/3212/3958

Jorge-Pozo, D., Gimenez-Gestal, C., Murillo, J. (2017). Influence of a virtual learning environment on Affectivity towards mathematics in high school students: a case study. Research in Mathematics Education XXI. Place: SEIEM.

Lafuente Martínez, M. (2003). Evaluación de los aprendizajes mediante herramientas TIC. Transparencies of assessment practices and pedagogical aid devices. [Doctoral thesis, [Doctoral thesis, Faculty of Psychology, University of Barcelona]. Dipòsit Digital de la Universitat de Barcelona.

Litwin, E. (1997): Teaching and classroom innovations for the new century, Bs. As. El Ateneo.

Medina, H., Lagunes Dominguez, A. (2020). What do Information and

Communication Technologies contribute to science education? Revista Digital Universitaria. (Vol. 21, No. 3). doi: http://doi.org/10.22201/codeic.16076079e.2020.v21n3.a9

Montoya A, Parra CMR, Lescay AM, et al. (2019). Pedagogical theories underpinning learning with the use of Information and Communication Technologies. RIC;98(2):241-255.

Morera, M. (2009). Electronic Manual: Introduction to Educational Technology. University of La Laguna. España.

Orjuela Forero, D. (2010). Integrating ICT to the curriculum in secondary education. UNAD Research Journal. Volume 09. Number 3. pp.137-156.

Partida Ibarra, J., Martínez García, M., Mena Hernández, E., Mercado Lozano, P., Pérez Zúñiga, (2018). Knowledge society and information society as the cornerstone in educational technological innovation. Iberoamerican Journal for Educational Research and Development. Vol. 8, No. 16.

Pérez Alarcón, S. (2010). The Importance of ICT in the School. Temas para la Educación. Number 7. (pp.1-7).

Plaza Gálvez, L., González Granada, Vasyunkina, O. (2020). Proposals for the Teaching of Mathematics. In Rebeca Flores (Ed.). ALME 33. (1 ed., Vol. 33, pp. 295-304). Latin American Committee of Educational Mathematics.

Recio Urdaneta, R., Recio Urdaneta, J., Saucedo Fernandez, M., Jimenez Izquierdo, S. (April 20-30, 2017). Connectivism, advantages and disadvantages. VII Congreso Virtual Iberoamericano de Calidad en Educacion Virtual y a Distancia.

Revelo-Rosero, J. and Carrillo Puga, S. E. (2018). Impact of the use of ICT as tools for the learning of mathematics of middle school students. Revista Cátedra, 1(1), 70-91.

Riveros V, Víctor S. and Mendoza, María Inés (2005). Theoretical bases for the use of ICT in Education. Educational Encounter. ISSN 1315-4079 ~ Legal deposit pp 199402ZU41 Vol. 12(3), pp.315 - 336.

Rodríguez, Daniela (May 13, 2020). ICT in everyday life: uses, advantages, disadvantages. Lifeder. Retrieved from https://www.lifeder.com/tic-vida-cotidiana/ .

Rubio Michavila, C., Pérez Martell, E. (1999). New educational models based on technologies. Revista Electrónica Universitaria de formación del Profesorado. 2(1).

Ruiz, G. (2020). Marks of the Pandemic: The Right to Education Affected. International Journal of Education for Social Justice, 2020, 9(3e), 4559. https://doi.org/10.15366/riejs2020.9.3.003

Salinas, J. (2013). Flexible Teaching and Open Learning, Key Fundamentals of PLEs. In L. Castañeda & J. Adell (Eds.), Personal Learning Environments: Keys to the networked educational ecosystem (pp. 53-70). Alcoy: Marfil.

Sanchez, Jaime H. (2002) Curriculum Integration of ICTs: Concepts and Ideas. Department of Computer Science, University of Chile.

Sobrino Morrás, Ángel Aportaciones del Conectivismo como modelo pedagógico post constructivist. Propuesta Educativa [online]. 2014, (42), 39-48[date of Consultation 6 of.
July 2021]. ISSN:. Available at:
https://www.redalyc.org/articulo.oa?id=403041713005

Terrazas Pastor, Rafael; Silva Murillo, Roxana (2013). Education and the knowledge society (Vol. 32, pp. 145-168). PERSPECTIVAS.

Torres Cañizález, Pablo César, Cobo Beltrán, John Kendry Tecnología educativa y su papel en el logro de los fines de la educación. Educere [online]. 2017, 21(68), 31-40[date of.
Accessed May 21, 2021]. ISSN: 1316-4910. Available at:
https://www.redalyc.org/articulo.oa?id=35652744004

Zapata-Ros, M. (2015). Theories and models on learning in connected and ubiquitous environments.
Bases for a new theoretical model based on a critical view of "connectivism", Education in the Knowledge Society (EKS).

7. ANNEX I

Teacher Survey:

<table>
<tr><td>

Subject:

School Year:

1. Do you implement ICT in the classroom and in what way?

2. Do you have the necessary resources to implement ICT in the classroom? If not, please indicate which ones.

3. How do students respond to the use of ICT?

4. Is any specific software taught that can be applied to different areas? Which ones?

</td></tr>
</table>

Student survey:

<table>
<tr><td>

Year of study:

1.   What do you know most about using technology?

Social Networking. Applications for the school.

2.   Do you use technologies in the classroom? Yes No

3.   What do you use most to work in class? Computer
Cell Phone Both

</td></tr>
</table>

4. When you are proposed to work with ICT:

I am enthusiastic I don't care I am not interested in what you are proposing to me
5. Do you use computer programs for mathematics? If yes, who teaches you how to use it and what programs do you use?

ANALYSIS:

In the surveyed courses of the Oriented Cycle (both 4th years), all of them use social networks and only one uses applications related to the school environment; in contrast to the basic cycle, enthusiasm for the proposed classes has increased, although there is still a high number of students who do not care about the proposals. It is observed that all students use cell phones and not computers.

In this case, the Mathematics teacher is the same and despite the fact that she uses the same resources in both courses, the students do not perceive the same thing. The 4th year students of the Physical Education course do not feel that they use mathematics programs when the teacher states that when graphing and arriving at conclusions they use the Geogebra software in the Android version for cell phones, as well as the use of different calculator programs according to the performance of the students in their devices. Likewise, the ICT teacher of this same course states that she implements the use of technologies as pedagogical resources, stating that she has the necessary tools to do so since the school has a good internet platform in operation and the students all have mobile devices. The software that the ICT teacher uses in this space are: Office package, Canva (Create in team), Kahoot (platform to create quizzes and contests in the classroom to learn or reinforce learning) and Clasroom (free educational platform from google). But at the same time he maintains that students are reluctant to implement them because they do not identify the use of ICT as a teaching tool and resource.

The same does not occur with the 4th Music Orientation, since the ICT teacher, although she incorporates computers so that students can "interact with them and their different programs: Word, Excel and Internet". She maintains that she is unable to use the "Mobile Computer Classroom" because there is a risk that other students will erase the work done.
Although she perceives enthusiasm from her students when she proposes them to work with ICT. (See graph).

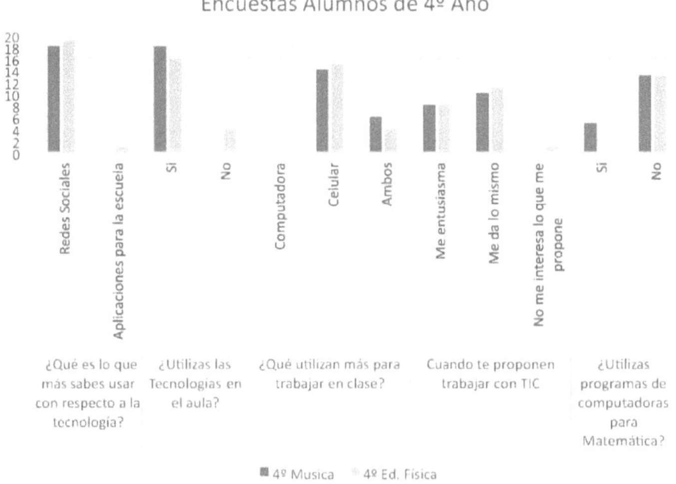

In all cases, the diagnosis carried out shows how students perceive what teachers believe to be novel and innovative for them, and in most cases what is proposed is not very attractive for the courses surveyed. It also shows the need for teachers to express what happens to them not only with ICTs themselves, but also with their implementation in the classroom. The ideal way to investigate teachers is through an interview and not through a survey, since in the latter they go beyond the proposed questions and in many cases they orally express their interest in working together to implement new strategies.

8. ANNEX II

Attendance to virtual classes in May and June 2021.

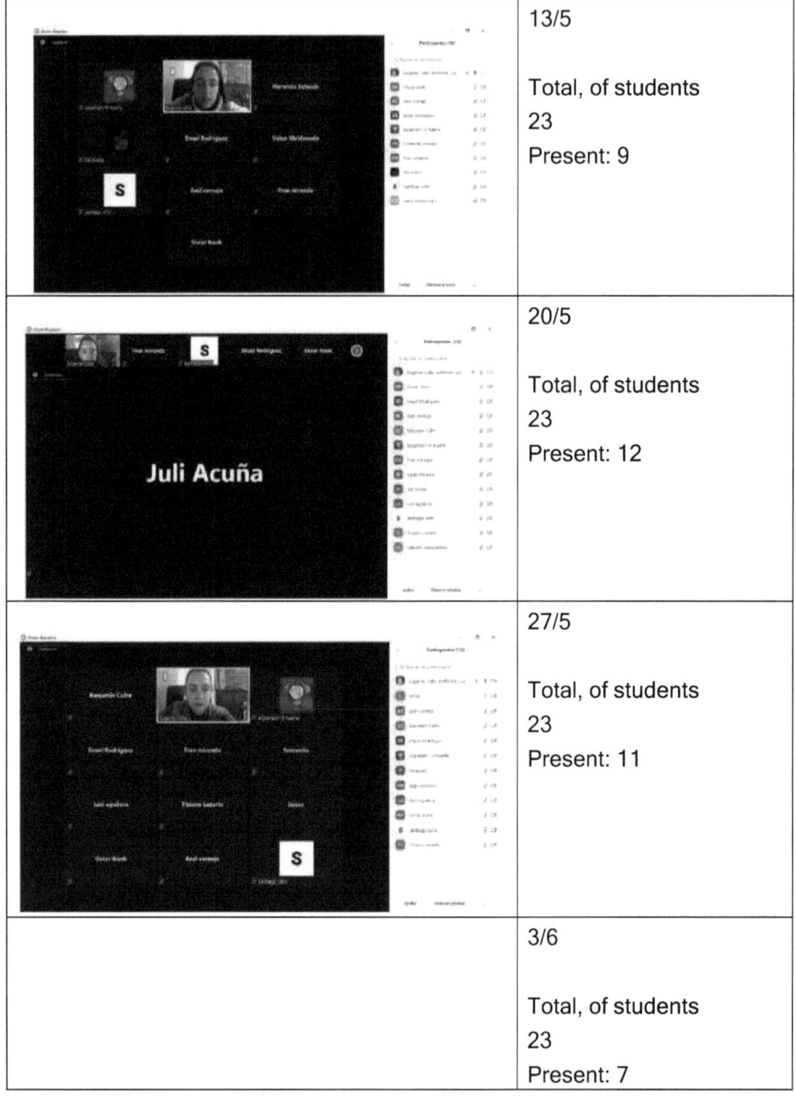

	13/5 Total, of students 23 Present: 9
	20/5 Total, of students 23 Present: 12
	27/5 Total, of students 23 Present: 11
	3/6 Total, of students 23 Present: 7

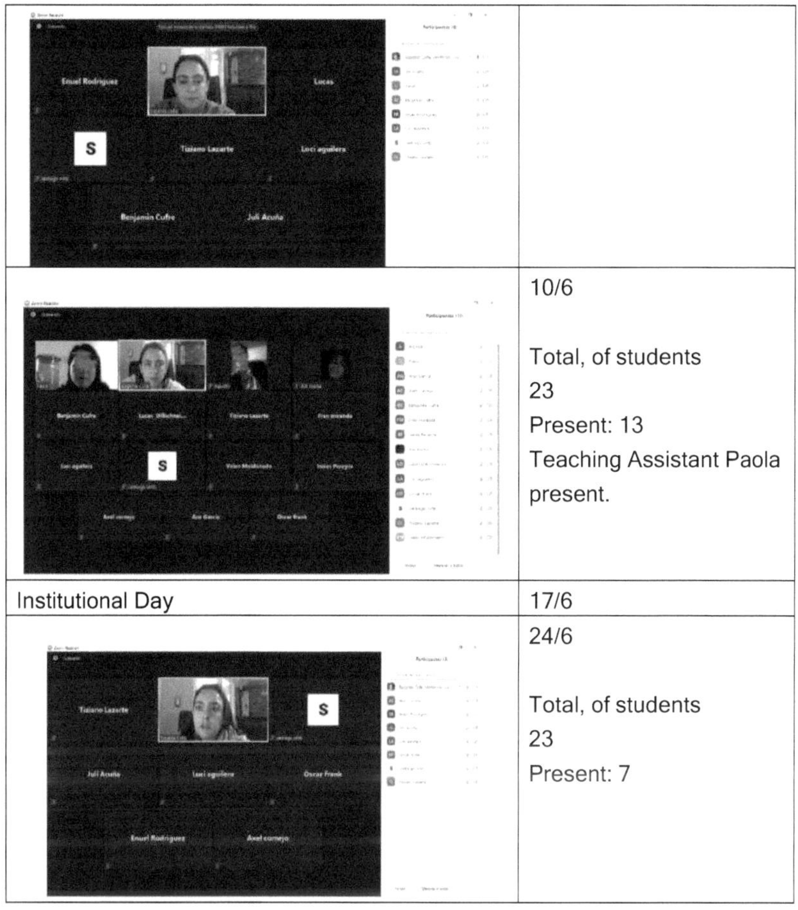

	10/6 Total, of students 23 Present: 13 Teaching Assistant Paola present.
Institutional Day	17/6
	24/6 Total, of students 23 Present: 7

I want morebooks!

Buy your books fast and straightforward online - at one of world's fastest growing online book stores! Environmentally sound due to Print-on-Demand technologies.

Buy your books online at
www.morebooks.shop

Kaufen Sie Ihre Bücher schnell und unkompliziert online – auf einer der am schnellsten wachsenden Buchhandelsplattformen weltweit! Dank Print-On-Demand umwelt- und ressourcenschonend produziert.

Bücher schneller online kaufen
www.morebooks.shop

Printed by Books on Demand GmbH, Norderstedt / Germany